Technology Intelligence

Technology Intelligence

Concept Design and Implementation in Technology-based SMEs

Pascal Savioz
Center for Enterprise Science
Technology and Innovation Management
Swiss Federal Institute of Technology (ETH)
Zurich

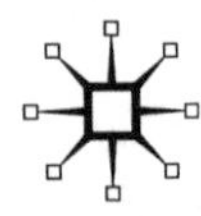

First published 2004 by
PALGRAVE MACMILLAN
Houndmills, Basingstoke, Hampshire RG21 6XS and
175 Fifth Avenue, New York, N.Y. 10010
Companies and representatives throughout the world

PALGRAVE MACMILLAN is the global academic imprint of the Palgrave Macmillan division of St. Martin's Press, LLC and of Palgrave Macmillan Ltd.
Macmillan® is a registered trademark in the United States, United Kingdom and other countries. Palgrave is a registered trademark in the European Union and other countries.

ISBN 1–4039–0583–5

This book is printed on paper suitable for recycling and made from fully managed and sustained forest sources.

A catalogue record for this book is available from the British Library.

Library of Congress Cataloging-in-Publication Data

Savioz, Pascal, 1973–
Technology intelligence : concept design and implementation in technology-based SMEs / Pascal Savioz.
p. cm.
Includes bibliographical references and index.
ISBN 1–4039–0583–5
1. Small Business–Management. 2. High technology industries–Management. 3. Technological innovations–Management. 4. Research, Industrial–Management. 5. Knowledge management. I. Title.

HD62.7.S266 2003
658.5′14–dc21

2003041426

10 9 8 7 6 5 4 3 2 1
13 12 11 10 09 08 07 06 05 04

Printed and bound in Great Britain by
Antony Rowe Ltd, Chippenham and Eastbourne

for Jean-Marc

Contents

List of Figures

List of Tables

Foreword

Hugo P. Tschirky

The existential dependency of technology-based enterprises on a continuous inflow of information from the technology environment has been an accepted reality since the emergence of technology management some decades ago. A new management issue, however, is the experience that the appropriate information supply no longer can be left to the sole initiative of single professionals but rather requires a systematized solution – often called technology intelligence – corresponding to the company-specific subjection to technological change. From a recent study on best-practice technology intelligence of large multinational enterprises, it becomes important that enormous efforts are made in order to establish structures and processes which aim at ensuring the vital inflow of relevant information.

The obvious question arises as to what measures should be taken by small and medium-sized companies which crucially rely on external information sources as well, but are notoriously limited in their resources. With this publication in hand Pascal Savioz is providing appropriate answers to this question. Based on perennial action research work and the management analysis of numerous high-tech companies, a concept of technology intelligence and its implementation is presented which meets the current needs and expectations of SMEs. In its core the concept originates on the *opportunity landscape*, a most useful metaphor for the purpose of sorting essential business knowledge in terms of its current, near-term, and future relevance. This structured outline of strategically significant knowledge fields provides a suitable basis for assigning individual technology intelligence responsibilities for each one of these fields. Such responsibilities include keeping abreast of scientific breakthroughs, upcoming product and process innovations, the activities of competitors, and periodic overall evaluation of the fields in terms of their strategic significance. Further key elements of the proposed technology intelligence system include the external expert network, the screening process, and the appropriate IT platform.

While this volume is a reference book for the design and implementation of technology intelligence systems, it also covers major issues of general technology and innovation management. In this regard the *technology management group* plays a central role. It comprises functions like elaborating technology strategies, handling patent and technology transfer issues, providing technical support to R&D and production, assuming responsibility for knowledge preservation, and managing the entire technology intelligence process.

As a whole this publication represents a remarkably well elaborated compendium for technology-based companies on ways to effectively cope with the challenges of technological change. Moreover it provides an excellent reference point for scholars *and* students in technology and innovation management who attach importance to dealing with management issues which are in fact field-tested.

Prof. Hugo P. Tschirky, Ph.D., DBA
Center for Enterprise Science
Technology and Innovation Management (ETH)
Swiss Federal Institute of Technology
Zurich

List of Abbreviations

ADIT	Agence pour la diffusion de l'information [information agency]
AG	Aktiengesellschaft [incorporation]
AO	Arbeitsgemeinschaft für Osseosynthesefragen (ASIF)
ASIF	Association for the Study of Internal Fixation
ASM	Association of Swiss Engineering Employers
BD&L	Business development and licensing
BI	Business intelligence
BioTech	Biotechnology
CAx	Computer-aided ... [design, manufacturing, engineering]
CEO	Chief executive officer
CI	Competitive intelligence
CIA	Central Intelligence Agency
CMG	Circle Member Group
CTI	Competitive technical intelligence
CTO	Chief technology officer
DB	Database
EC	European Community
EIRMA	European Industrial Research Management Association
EITM	European Institute of Technology Management
ETH	Eidgenössische Technische Hochschule [Swiss Federal Institute of Technology]
EU	European Union
FSRM	Swiss Foundation for Research in Microtechnology
h.c.	*honoris causa*
HCI	Healthcare industry
HRM	Human resource management
IS	Information system
ICT	Information and communication technology
IP	Intellectual property
ISIC	International standard industrial classification
IT	Information technology
ITI	International Team for Oral Implantology
IVD	In-vitro diagnostics
KTI	Kommission für Technologie und Innovation [Commission for Technology and Innovation]

MedTech	Medical technology
Mio	Million
MEP	Manufacturing extension partnership
NSA	National Security Agency
OECD	Organization for Economic Cooperation and Development
OEM	Original equipment manufacturer
QM	Quality management
RAB	Research advisory board
RD	Roche Diagnostics
R&D	Research and development
R&T	Research and technology
RIC	Roche Instrument Center
RMB	Research management board
SBA	Small business administration
SCIP	Society of Competitive Intelligence Professionals
SCM	Supply-chain management
SEM	Scanning electron microscope
SLA	Sand-blasted, large grit, acid-etched
SMEs	Small and medium-sized companies
S&T	Science and technology
TAB	Technology advisory board
TI	Technology Intelligence
T&E	Training and education
TF	Technology field; technology forecasting
TMG	Technology management group
TPM	Technology platform management
TRM	Technology road-mapping
US	United States
VC	Venture capital
VSM	Swiss Association of Machinery Manufacturers

1 Introduction

Focus of the book and main questions

> Accelerated technological change has become a fact and will continue to challenge industrial and societal development in this new century.
>
> Tschirky (2002: 1)

Anticipating these changes seems to be crucial for success in technology-based companies. Being informed about the company's technological environment in a time of rapid change is prerequisite to a conscious management of technologies. Cohen and Levinthal (1990: 128) noted that the most innovative firms appear to be those that are best at recognizing the relevance of new, external information, importing and assimilating it, and then applying the information. However, companies are facing more and more problems getting and handling this information because of the increased complexity of the technological environment (Abernathy & Utterback 1978: 40). There are multiple causes and effects of this increasing complexity; to name a few:

- **The phenomenon of technology fusion**: The interdisciplinarity of scientific fields is increasing. Thus, a technology cannot be assigned to one specific science. As a consequence, barriers between sciences become blurred: a technology field combines several sciences (for example, the emergence of biotechnology can be interpreted as a triangular connection between food, drugs and medicine, and industrial chemicals). Kodama (1991: 130) introduced the expression 'technology fusion' for this phenomenon.

- **The global scope of technologies:** The worldwide integration and coordination of R&D is a principal problem of companies. The challenge is to optimize internal and external knowledge transfer (Von Zedtwitz, 1999: 8).
- **The tremendous explosion of technical knowledge production:** Every day over 6,000 scientific publications are released. Every 10 or 15 years the amount of published scientific literature doubles. These numbers have a human equivalent: today over 5 million people work in the area of knowledge production in R&D departments; this is approximately 90 percent of all scientists who have ever lived (Nefiodow 1990: 50).
- **Longer development times coping with shorter market cycles:** The increasing technological complexity requires longer development times, and the diversity of innovations causes shorter market cycles, which means that the time of cash outflow increases while the time of cash in-flow decreases (Tschirky 1998a: 5).
- **Escalating costs of internal R&D:** In the pharmaceutical industry, the development costs of a new drug amount to at least US$350 million with a development time of about 10 years (Bethke & Lang 1998: 701).

In order to cope with the challenge of being informed about the technological environment, the management literature and numerous companies have developed a broader interest in the field of **Technology Intelligence (TI)**.

The emerging interest in the "intelligence discipline" can be illustrated by the growth in membership of the Society of Competitive Intelligence Professionals (SCIP) over the last 15 years. Since 1997, SCIP has organized an annual symposium on Competitive Technical Intelligence.

The breakthrough in the practical application of intelligence in industry, however, seems to lag. While most companies answered in a survey that they were acting in a highly competitive market, just a few companies were applying intelligence techniques systematically, or running a regular intelligence system. The survey pointed out managers' need for intelligence activities, especially in respect to technology and their company's direct competitors. On a 10-point scale, general technology (8.8), industry-specific technology (9.0), and new products (9.0) were classed as "things that companies will need to keep track of to stay competitive" (Hall 2001: 5).

There are some studies which report on the Technology Intelligence current in large, multinational companies – for example, Lichtenthaler (2000). In turn, from both theoretical and practical points of view, two questions in the field of TI still remain unanswered:

- How do small and medium-sized enterprises (SMEs) handle the topic of "Technology Intelligence"? There is little insight into how SMEs are organized to identify, collect, analyze, and apply relevant information about trends in the company's technological environment. It is not clear whether insights from concepts applicable to large organizations can be transferred to SMEs.
- How can Technology Intelligence concepts be implemented? In addition to conceptual work in the field, studies have been conducted on this topic which examine the state of the art of processes and methods of TI through case studies. Examining implementation seems to be highly interesting to companies.

This book aims to find answers to these open questions. Thus:

The topic of this book is Technology Intelligence, and the considered objects are technology-based small and medium-sized enterprises (SMEs).

In order to be able to handle the topic, the formulation of the **main questions** is:

How could a Technology Intelligence system be designed for a technology-based SME?

How could a technology-based SME proceed to implement a Technology Intelligence system?

Over the last couple of decades the interest in TI has shifted from an academic to a pragmatic debate. The emphasis of this book is on gaining insight into business reality by means of first-hand information in order to be able to develop solutions that are of use to practitioners. Thus, **the main output is normative** rather than a simple description of "Technology Intelligence in use in technology-based SMEs."

This book is based on academic research. However, practical illustrations are presented for the practitioners' needs. Therefore, the author has in some instances attempted to minimize theoretical discussion.

Procedure and structure of the book

The **knowledge creation procedure in this book** is depicted in Figure 1.1. In fact, the main questions as well as answers to them are based on close interaction between industrial reality and a sound theoretical basis.

The book is structured in nine principal chapters:

Chapter 1 sets the research focus and formulates the main questions in order to outline the content of this book.

Chapter 2 principally defines terms and important issues. However, it goes beyond a simple definition chapter, to explore terms such as "technology-based" and "SMEs," the "information–knowledge–intelligence" interaction, and major management issues with regard to TI, such as "strategic management," "technology management," "innovation management," and "knowledge management."

Chapter 3 clarifies the relevance of the topic in practice. It is not the goal to examine practices in companies but to examine interest in the topic. Thus, this chapter is not solution-oriented, but shows problems and needs of SMEs' reality. The practitioner's voice is captured by means of interviews, through seminars and workshops.

Chapter 4 illustrates the state-of-the-art of Technology Intelligence in theory and builds the theoretical basis of this book. The goal of this chapter is twofold: first, the chapter aims to illustrate the gap for TI in technology-based small and medium-sized enterprises from a theoretical point of view. Secondly, analysis of recent literature in the field of (general) TI and related topics (for example business and competitive intelligence) will build the theoretical basis for this book. The notion of the Technology Intelligence System will be introduced in this chapter. This system is inspired by Michael Porter's (1985) value chain: The intelligence process is a value-creating process, including the steps of need identification, information collection, analysis, communication of insight, and intelligence use. The value added is the improvement of decision-making quality. Enablers of these direct intelligence activities are TI Mission and Goals, TI Structures, TI Tools, and Management of the TI System.

Chapter 5 provides a résumé of insight into the topic and sets the design of the research part of this book. The research in this book is practice oriented and therefore easily readable and of interest to the practitioner. A two-step procedure is chosen: action research (Chapter 6), in order to build knowledge about how to design and implement a TI system in a technology-based SME; followed by a validation of the generated knowledge (Chapter 7).

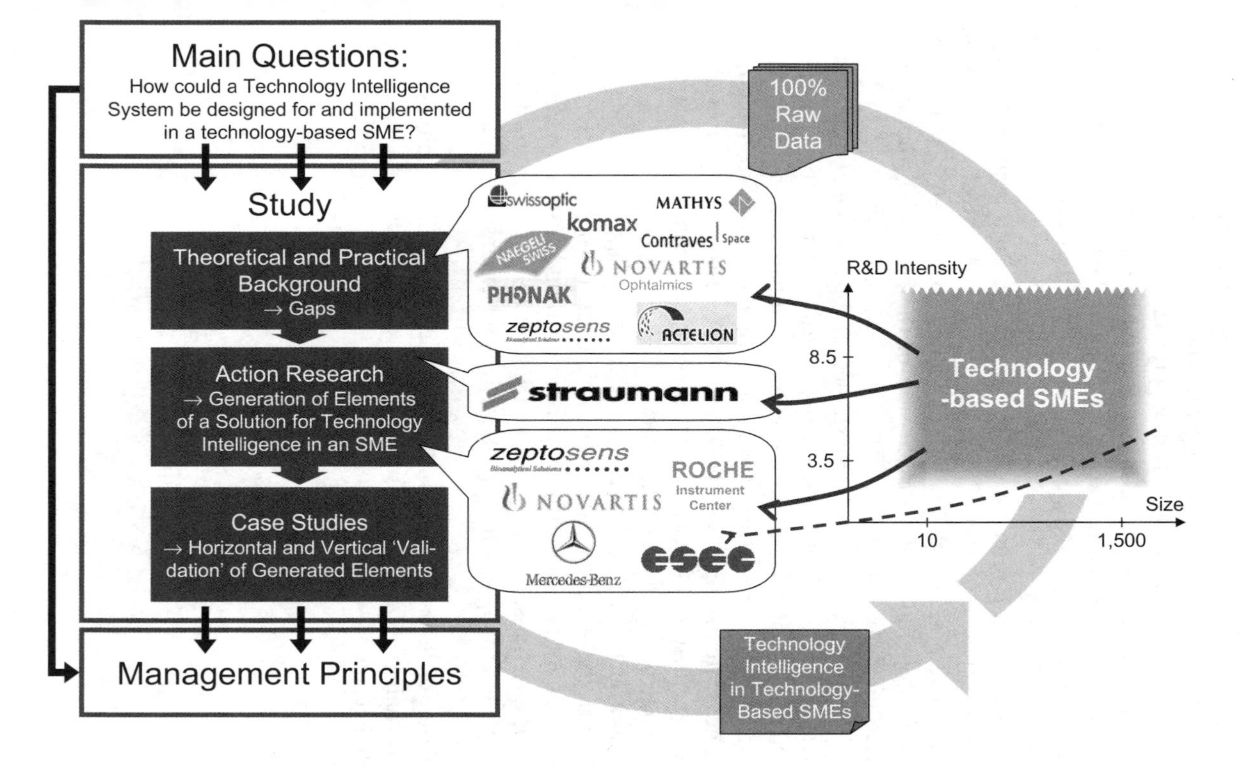

Figure 1.1 The knowledge creation procedure in this book

Chapter 6 contains an in-depth action research case conducted at a medium-sized Swiss company in the field of medical technology. The aim of this chapter is to find, together with practitioners, a solution for how technology-based SMEs can design and implement a TI system. In order to understand the context and the conditions in which this action research is embedded, a case study of the initial situation starts this chapter. Then insight is summarized and action requirements are formulated. This is the divide between the non-influenced and influenced action research environment. The third (and major) part aims to generate elements for a solution for TI in technology-based SMEs. These elements will be explained extensively. Finally, this chapter concludes with a discussion of the generated solution with regard to practice and theory.

Chapter 7 presents some case studies for the purpose of validation of the elements found during the action research. In fact, this chapter is a mirror of Chapter 5. Two case studies in companies comparable to the action research company discuss and validate the core elements of the generated solution for TI. Then, two case studies aim to validate the overall solution with regard to very small companies (start-ups) and large multinational companies.

Chapter 8 points towards a new set of management principles in order to make a practical conclusion. Thus, in the discussion of the management principles the practitioner will find answers to the initial major questions. In total, 10 management principles are formulated. For each principle some key benefits and extensive explanation is added.

Chapter 9, finally, concludes with a short summary and sketches new challenges and issues of further interest in the field of Technology Intelligence.

2
Definitions and Management Issues

The terms "intelligence" and "technology" are closely related to the term "knowledge." However, a unique definition of the term knowledge does not exist. Depending on the context and the background, practitioners and scientists employ their own definitions. In order to get a better understanding of the terms knowledge, intelligence, and technology, and how they are used hereafter, different aspects are presented in the following.

Knowledge, information, data: definitions and interaction

According to Davis and Botkin (1994: 166), **data** comes to us in four different forms: numbers, words, sounds, and images. They are worthless until they are related in a particular context. **Information** is analyzed data that has been arranged into meaningful, object-oriented patterns. The value and usefulness of information depends on the absorptive capacity (the ability to evaluate and utilize outside knowledge) of the recipient (Cohen & Levinthal 1990: 132). Once the recipient assimilates, interprets, evaluates, and uses information, we can talk about **knowledge** (Koruna 2001: 100). In this book, knowledge is understood as the totality of experience, cognition, and skills of individuals to solve a problem (Probst et al. 1999: 46). Thus, knowledge is always action-oriented and personal. Speaking about the knowledge of companies, Probst et al. (1999: 46) define "the **organizational knowledge base** as the totality of individual and collective experience, cognition and skills to which the organization can access in order to solve a problem, including all underlying data and information."

The transfer and interaction of data, information, and knowledge is shown in Figure 2.1. If one accepts this model, a direct and formal transfer of knowledge from one individual to another individual, and therefore to the company, is not possible. In fact, parts of knowledge should be abstracted and decoded in order to be transferred at a lower level. This mechanism of interaction should be understood as a continuum between the different levels (Probst et al. 1999: 39).

Since the receiver has to reinterpret the information in turn, a part of the original significance gets lost (Boisot 1983: 165; 1998: 14). Furthermore, Polanyi (1966: 4) argues that just a part of knowledge can be articulated explicit knowledge, and therefore be transferred as shown in Figure 2.1. Implicit knowledge, on the other hand, can be neither easily articulated nor transferred via information or data, but can only be transferred through direct interaction between individuals via metaphors and analogies. Nonaka and Takeuchi (1995: 62) adopt this epistemology (theory of cognition) and describe in detail four modes of how knowledge can be converted: from tacit to tacit (socialization), from tacit to explicit (externalization), from explicit to explicit (combination), and from explicit to tacit (internalization). In addition, they argue that an iterative interaction between tacit and explicit knowledge (the knowledge spiral) becomes larger in scale as it moves up the ontological levels from an individual to a higher level. In its proper definition, knowledge is created by one individual. However, the organization supports this knowledge creation, which therefore should be understood as process (Nonaka & Takeuchi 1995: 74).

This evolution of knowledge and knowledge creation enriches the organizational knowledge base in quality and quantity, and therefore could be interpreted as **organizational learning** (Probst et al. 1999: 46). The effectiveness and efficiency of organizational learning depends on the absorptive capacity of the firm (Cohen & Levinthal 1990: 131), on the company culture (Cook & Yanow 1996: 448), and on the capability of the company to apply insight from mature to new concepts (Senge 1990: 174). Junnarkar (1997: 32) argues that the company's challenge is to be able to take the learning from successful, mature concepts and apply it to other mature concepts. He calls this adaptive learning or single-loop learning. Concurrently, organizations need to foster generative learning, or double-loop learning. This will be manifested by the number of new concepts which are born in the organization and are nurtured. Moreover, he argues that these different types of learning need different skills within the organization. The consequence for organizational learning is that the path from generative to mature concepts needs both, more complete information and a higher clarity of understanding (Figure 2.2).

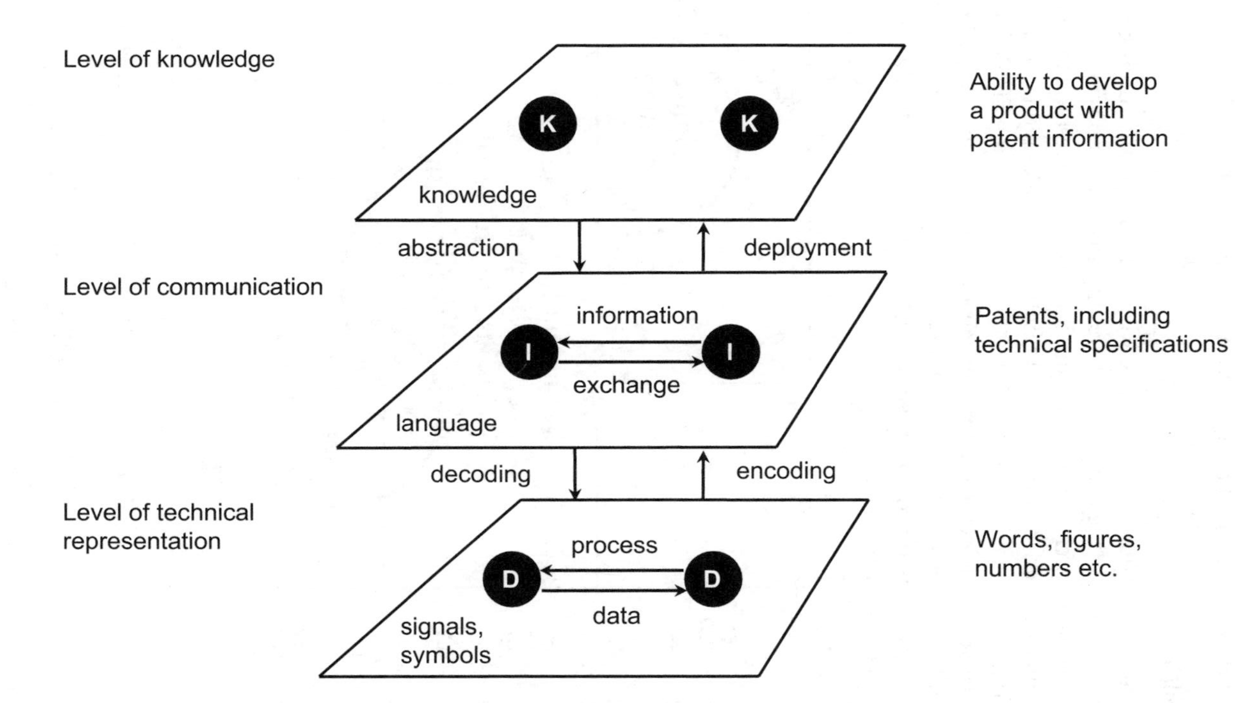

Figure 2.1 Interaction and transfer of data, information, and knowledge

In conclusion, in this book knowledge is accepted as a company's most important resource; and organizational learning is the primary condition for potential competitive advantage.

Intelligence

There is a very fuzzy distinction between knowledge and **intelligence**. While some authors understand intelligence as synonymous with knowledge, other authors see intelligence as something between information and knowledge. Fuld (1995: 23) speaks about "analyzed information," Prescott and Smith (1989: 164) state that "intelligence is information whose credibility and meaning have been established," Bryant et al. (1997: 159) emphasize an action relationship: "Intelligence is information which has been analyzed to the extent that decisions can be made."

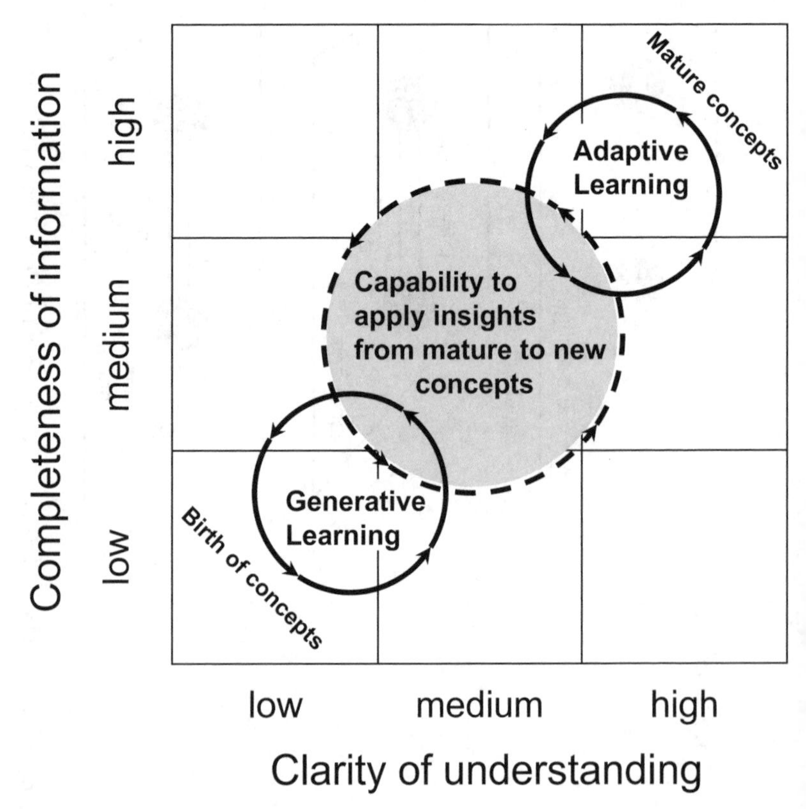

Figure 2.2 Adaptive learning versus generative learning

In this book, the term intelligence is understood as "information which has been analyzed in order for the best possible decision to be made." Thus, intelligence on its own is not action related (contrary to knowledge), but builds the basis for taking action, i.e. a decision. The depth of clarity of understanding and the completeness of information required depend on the context. Furthermore, unlike knowledge, intelligence is not personal. This interaction between information, knowledge and intelligence is represented in Figure 3.3.

Technology

According to Tschirky (1998b: 227), **technology** means "specific individual and collective capabilities in explicit and implicit forms for the product and process oriented deployment of scientific and engineering knowledge." On the one hand, technology can be understood as a subset of knowledge (as defined below), and the flow of genesis of technologies is the same as for knowledge. On the other hand, technology might be appreciated as an output of the process of the knowledge deployment.

In this book, this broad view of the term "technology" is fully accepted. As a consequence, the term can be used at different levels of transformation during its evolution (for example, different steps in microchip research and development) and different levels of aggregation (for example, nanotechnology and surface technology).

In relation to intelligence, it makes sense to precisely describe the types of media of technologies. Ewald (1989: 40) differentiates between the three types: personal media (for example, researcher, manager, and lawyer), informational media (for example, patents, licenses, blueprints, software, databases, documents), and material media (for example, instruments, prototypes, products).

An attempt to define technology-based SMEs

"A small business is not a little big business!" (Welsh & White 1980). This sentence, still valid today, shows the problem of a large number of companies. While an enormous percentage of companies are in fact "small" businesses, most theories are developed for "big" businesses. One kind of theory is not readily transferable to the other, mainly because small businesses have restricted resources (in addition to traditional resources, such as labor, capital, and property, the "most important resource," knowledge). "Compared with larger companies, SMEs suffer from a lack of resources and personnel. Multiple roles being filled

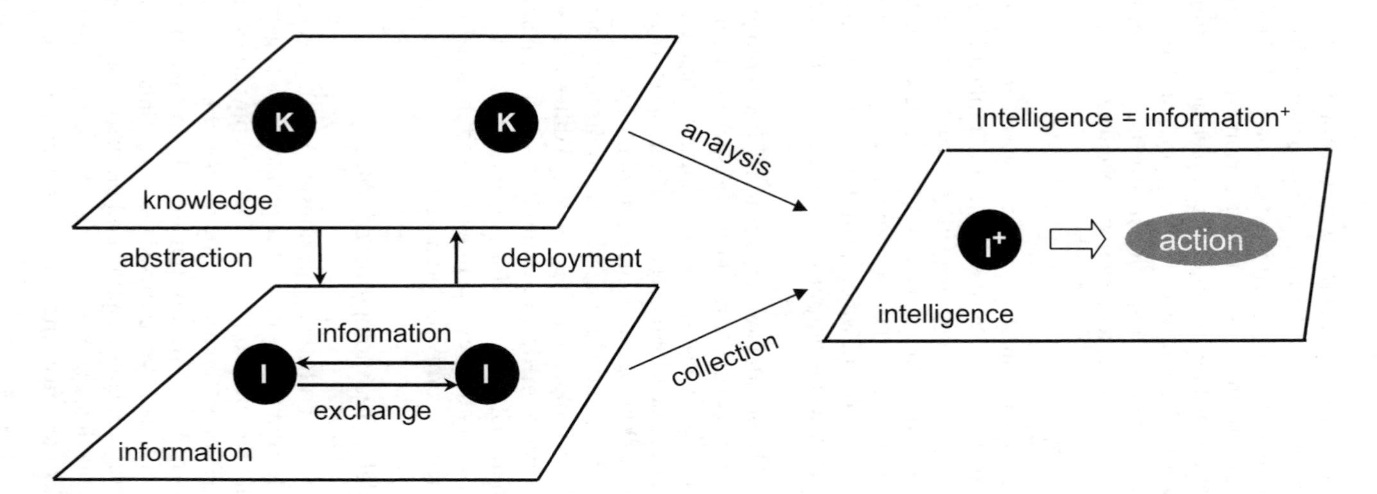

Figure 2.3 The interaction between information, knowledge, and intelligence

by staff and a lack of organizational slack make dealing with anyone outside of the company much harder. They are disadvantaged by nature when it comes to accessing new ideas" (Major & Cordey-Hayes 2000: 589). Many small firms not only lack resources, but probably also lack the relevant specialized and co-specialized assets (Hadjimanolis 2000: 267).

On the other hand, recent research on "small" businesses focused on small, high-tech start-ups, i.e. the Silicon Valley type of firm (Cobbenhagen 2000: 34). Small and medium-sized enterprises are hardly covered at all. Yet most of the growth in employment and added value can be attributed to them. But what are SMEs? How can they be defined?

Some existing definitions of SMEs

Literature on the definition of SMEs differentiates between quantitative and qualitative criteria. Quantitative criteria range from size in terms of employees or turnover, to market shares. Because of its practical aspect, the number of employees and the company's turnover seem to be the most appropriate way to define SMEs. However, a homogenous definition does not exist. Table 2.1 gives an overview of different quantitative definitions in the USA, Japan, Europe overall, Germany, and Switzerland. Some authors propose specific definitions in different industrial sectors because of their different characteristics. However, this is not generally accepted and is in conflict with standard definitions.

An international comparison of a distribution of manufacturing enterprises by sizes (numbers of employees) is given in Figure 2.4. Following this figure, most enterprises are SMEs, independent from a precise definition. An estimation of some other main indicators of nonprimary private enterprises in the European Union (EU) is given in Table 2.2.

One advantage of "black-and-white" definitions – for example by numbers of employees – is clear limits. This is particularly necessary for tax advantages, and national or international promotional programs (EU Commission 1999). However, since definitions differ considerably (limits between 250 and 1,500 employees), there is no generally accepted rule. In addition, clear limits always consider the company as a "black box." Such a definition does not take into account the organization and the "inner life" of a company, and is therefore not the best definition for any research project. Is a company that has sales departments overseas (i.e. no R&D or production in subsidiaries) "bigger"

Table 2.1 Quantitative definitions of SMEs (by no. of employees and turnover)

USA	*Japan*	*Europe*	*Germany*	*Switzerland*
Different size standards for different industries *SMEs*: up to 500 employees for most manufacturing industries, US$17 million of turnover Many exceptions exist: one-fourth of SBA subsidized SMEs had varying sizes ranging from 500 to 1,500 employees!	Different size standards for different industries *SMEs*: up to 300 employees for most manufacturing industries, US$2.5 million of turnover	*Microenterprises*: 1–9 employees, (no statement about turnover) *Small enterprises*: 10–49 employees, max. US$3 million of turnover *Medium-sized enterprises*: 50–249 employees max. US$17 million of turnover *Large*: 250 and more employees, more than US$17 million of turnover	*Small enterprises*: up to 9 employees, annual turnover up to US$0.43 million *Medium-sized enterprises*: 10 to 499 employees, annual turnover US$0.43 to 43 million *Large*: 500 or more employees, annual turnover US$43 million and more	*SMEs*: less than 250 employees and max. US$34 of turnover

than a company that has distribution partners? Qualitative criteria may help to strengthen the understanding of a specific group of SMEs. Some criteria are (Clemens et al. 1997: 2):

- Identity of ownership and personal responsibility for the enterprise's activities,
- Identity of ownership and personal liability for the entrepreneur's and the enterprise's financial situation,
- Personal responsibility for the enterprise's success or failure,
- A personal relationship between employer and employees.

These criteria can be summarized as strong involvement of the owner (a single person or a family) in decision-making, and that the owner holds a majority of shares. This implies that SMEs may in fact employ more than 500 employees (Clemens et al. 1997: 3 and Hauser 2000: 2).

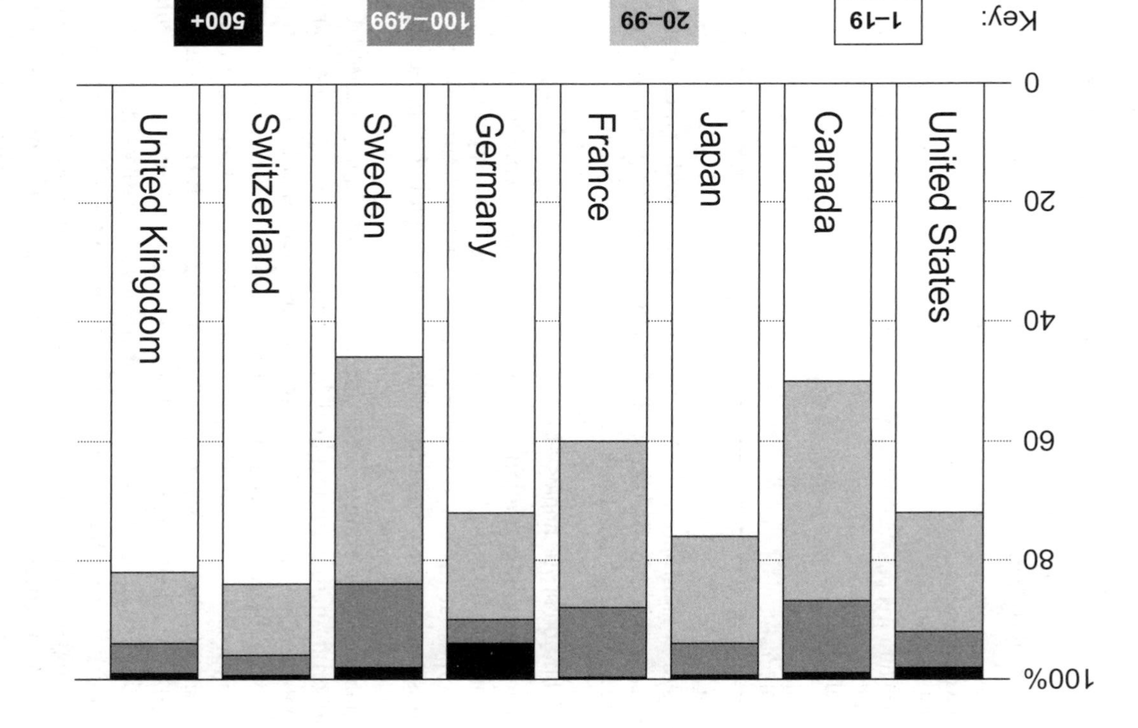

Figure 2.4 Distribution of manufacturing enterprises by size (number of employees) (OECD 1997: 72)

Table 2.2 Main indicators of nonprimary private enterprises in Europe (adapted from EU Commission 2000: 4, 5)

Indicator	*SMEs*	*Large enterprises*	*Total*
Number of companies [1,000s]	19,810	40	19,850
Employment [1,000s]	77,100	39,860	116,970
Development of employment over the last 10 years	2.5%	4%	—
Average turnover per enterprise [US$ millions]	0.44	188.6	0.877
Total turnover [US$ billions]	8,690	7,545	16,236

This discussion shows the heterogeneity of the existing quantitative and qualitative definition criteria. Therefore, an existing definition will not be adopted for this book.

The definition of "technology-based"

Management literature often uses terms like "technology-based," "technology-intensive," and "high technology" synonymously. It is surprising that there is not a generally accepted definition for these expressions. Most contributions that are about technology-based firms do not define the term.

Chabot (1995) examines the use of the term "high technology." Based on numerous authors he differentiates between input-based and output-based definitions. Two major factors drive input-based analyses: R&D expenditure and occupational profile statistics. The advantage of these approaches, if proper data is available, is that high-technology analysis is fairly straightforward. By counting gross R&D expenditures in dollars or calculating the number of technical staff, it is not difficult to arrive at an ordered spectrum of technology-intensive companies. It remains only to select a certain percentage to set a limit separating technology-based and non-technology-based companies. One example for R&D expenditures could be the OECD classification. The limit between low-technology and high-technology is 3.5 percent, the limit between high-technology and leading-edge technology is 8.5 percent. In contrast to a definition at the company level, a precise definition at the industry level is tricky. Since, sometimes, one company cannot be attributed clearly to one single industry, the popular understanding of "industries" differs from official industrial classifications (for example the ISIC). Figure 2.5 gives some examples of high-technology and low-technology industries following the OECD definition (OECD 1997: 110), and some examples of typical high-technology industries in the popular understanding.

An example of an occupation profile is Beyers and Nelson's (1998: 4) definition: technology-based industries employ 10 percent or more of their staff in R&D functions. The disadvantage of such input-based factors is that they neglect technology use and production. Output-based definitions are those which classify high technology based on the productive value-added output of firms. They are less common than the input-based methodologies. The advantage of defining it by its "sophistication," "complexity," or "advanced technology" content is that the actual products of intense R&D, rather than dollar input, drive the essential meaning of "high technology." There are important disadvantages to the output-based approaches, however, and they explain in part the relative abundance of input-based methods. First, output-based definitions rely on data that is neither highly accessible nor easily processed. Their second primary disadvantage is the high degree of subjectivity (Chabot 1995: 8).

An alternative definition would be Kodama's (1991: 49) model of more or less technology-intensive sectors. He defines industries by the cancellation rate in relation to the investment level of R&D programs. While the cancellation rate decreases the more the company spends on an R&D program in conventional sectors, the cancellation rate in science-based industries remains constant. The high-technology sector's behavior is the same as for the conventional sector up to a certain investment level. Then, it remains constant as for the science-based sector.

Another interesting view comes from Dankbaar (1996: 103). Within the totality of companies he differentiates between technology-intensive and technology-contingent enterprises. While the first group anticipates technological change, the latter treat technology as a contingency which appears by surprise and needs to be dealt with if it cannot be avoided. This view gives us two major insights: firstly, for every company technology – more precisely, technological change – is of concern. No company can escape this fact. As a consequence, every company should be aware of this, and should take appropriate measures. Secondly, as Dankbaar's study shows, there apparently are

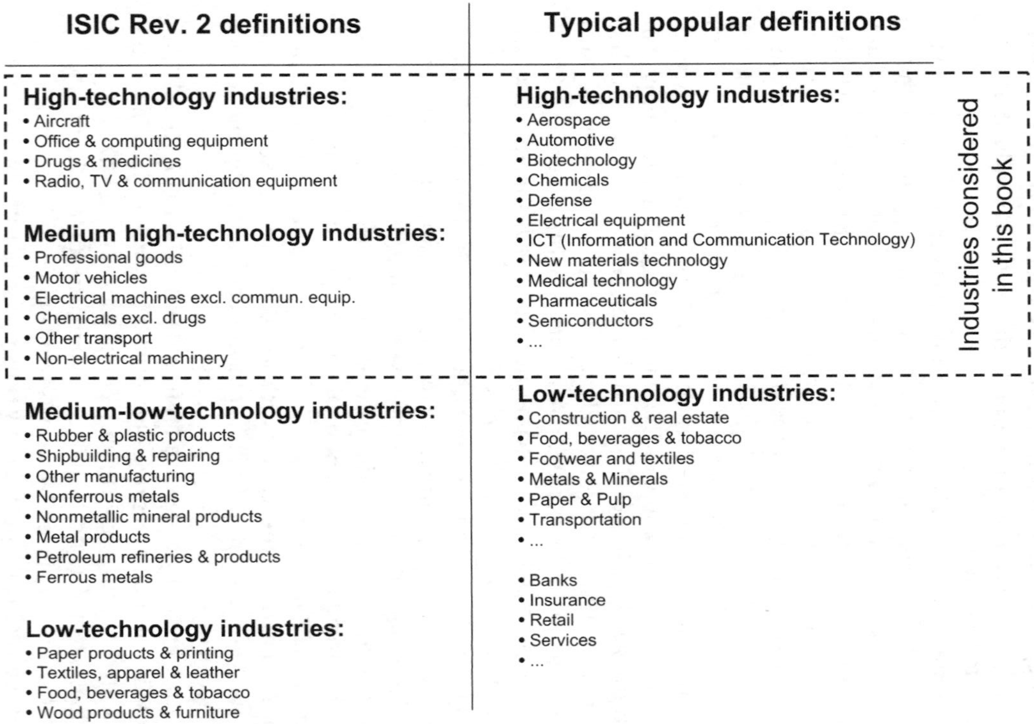

ISIC Rev. 2 definitions

High-technology industries:
- Aircraft
- Office & computing equipment
- Drugs & medicines
- Radio, TV & communication equipment

Medium high-technology industries:
- Professional goods
- Motor vehicles
- Electrical machines excl. commun. equip.
- Chemicals excl. drugs
- Other transport
- Non-electrical machinery

Medium-low-technology industries:
- Rubber & plastic products
- Shipbuilding & repairing
- Other manufacturing
- Nonferrous metals
- Nonmetallic mineral products
- Metal products
- Petroleum refineries & products
- Ferrous metals

Low-technology industries:
- Paper products & printing
- Textiles, apparel & leather
- Food, beverages & tobacco
- Wood products & furniture

Typical popular definitions

High-technology industries:
- Aerospace
- Automotive
- Biotechnology
- Chemicals
- Defense
- Electrical equipment
- ICT (Information and Communication Technology)
- New materials technology
- Medical technology
- Pharmaceuticals
- Semiconductors
- ...

Low-technology industries:
- Construction & real estate
- Food, beverages & tobacco
- Footwear and textiles
- Metals & Minerals
- Paper & Pulp
- Transportation
- ...

- Banks
- Insurance
- Retail
- Services
- ...

Industries considered in this book

Figure 2.5 Examples of high-technology and low-technology industries

companies which anticipate technological change better than others. Dankbaar questions whether technology-contingent enterprises will be able to survive in a fast-changing technological environment. In order to survive, he says, it is crucial to learn to monitor technological developments, and to react quickly to relevant changes.

The definition of technology-based SMEs used in this book

In order to define and justify a certain sample of enterprises, limits have to be set for both degree of technology intensity and the dimensions of the companies. The discussion about quantitative and qualitative criteria for company characteristics showed that a generally accepted definition does not exist, and that limits have to be set in relation to the focus of interest. Despite some arguments against quantitative criteria, this book uses the number of employees to delimit the size of the company. The main argument is that defining an SME by means of its size is very popular and is easily understood. In addition, it is simple to measure. Since the research focus is on the organization of Technology Intelligence (TI), which implies aspects like different roles and organizational learning, the company should have a certain number of employees. In this book, a lower limit (10 employees) is imposed to exclude microenterprises. This excludes small start-ups, due to the fact that a wide gap in literature exists for companies that fall in-between start-ups and large companies (Cobbenhagen 2000: 34). The upper limit is based on the highest existing definition (1,500 employees). However, both limits are not absolutely clear. Qualitative criteria do not enter the definition explicitly. But in this book, SMEs are nevertheless understood as rather independent from a large group, with a strong identification with the company among employees, and with a manifest and active owner.

As mentioned above, there is no generally accepted definition for terms "technology-based" or for "technology-intensive." Just a single criterion, R&D expenditure as a percentage of turnover, defines technology intensity in this book, again because it is commonly accepted and it is easy to determine. The OECD limit of 3.5 percent seems to be an appropriate lower limit for technology intensity. Again, this limit is not sharp. There is no necessity to define an upper limit: the more the company spends on R&D, the more technology is of top concern, and the more knowledge about technological change becomes relevant. The **definition of technology-based small and medium-sized enterprises** considered valid in this book is represented in Figure 2.6.

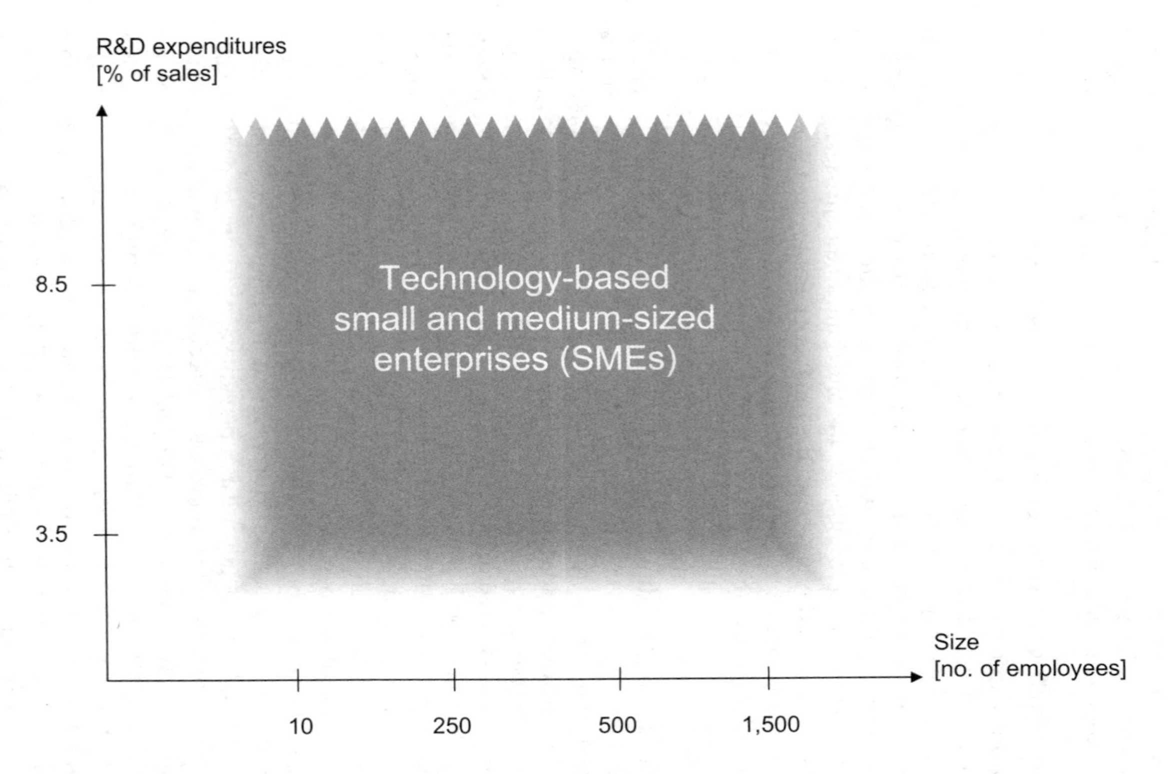

Figure 2.6 The definition of technology-based SMEs employed in this book

For example, in Switzerland, this size definition (10 to 1,500 employees) covered 62 percent of employment in 1998, compared to 16 percent for enterprises larger than 1,500 employees, and 22 percent for microenterprises. In terms of companies, 47,765 companies fall into this definition. There is no data available to give a sound scientific estimation of how many employees/companies within this size definition spend more than 3.5 percent for R&D expenditures. Beyers and Nelson (1998: 9) report that about 10.4 percent of employment in Washington State is in technology-based industries. Thus, a guesstimate lets us assume that about 6.5 percent (10.4% of 62%) of employment in Switzerland is in technology-based SMEs. Considering the number of companies, Arvanitis (1997: 83) found that about 38 percent of companies in Switzerland spend more than 3.5 percent for R&D. This makes about 18,000 (38% of 47,500) technology-based SMEs in Switzerland. These numbers are illustrated in Figure 2.7.

Management issues in technology-based SMEs

Various standard works on SMEs, especially in relation to innovation, include contributions on different management issues which are of importance to the focus of this book. From an initial point of view, four management issues seem to cover the basics with regard to Technology Intelligence: strategic, technology, innovation, and knowledge management (see Figure 2.8). The aim of this section is to briefly introduce each of these issues and to discuss them with regard to SMEs.

Strategic management

The strategic management issue is probably the most philosophical topic in management literature. As will be shown later, there are diverse schools and approaches to dealing with the strategy question. These schools emphasize that in recent decades the understanding of strategy has been undergoing a great change, certainly based on the insight that the future is less planable than expected. From an initial point of view, strategic management is understood as planning to run and change the business in order to achieve the business mission and goals (Wright et al. 1992: 3). Terms like "thinking," "acting," and "decision-making" are central to this purpose (Gälweiler 1987: 65).

Mintzberg and Lampel (1999) discuss 10 different schools of strategy formation, here summarized:

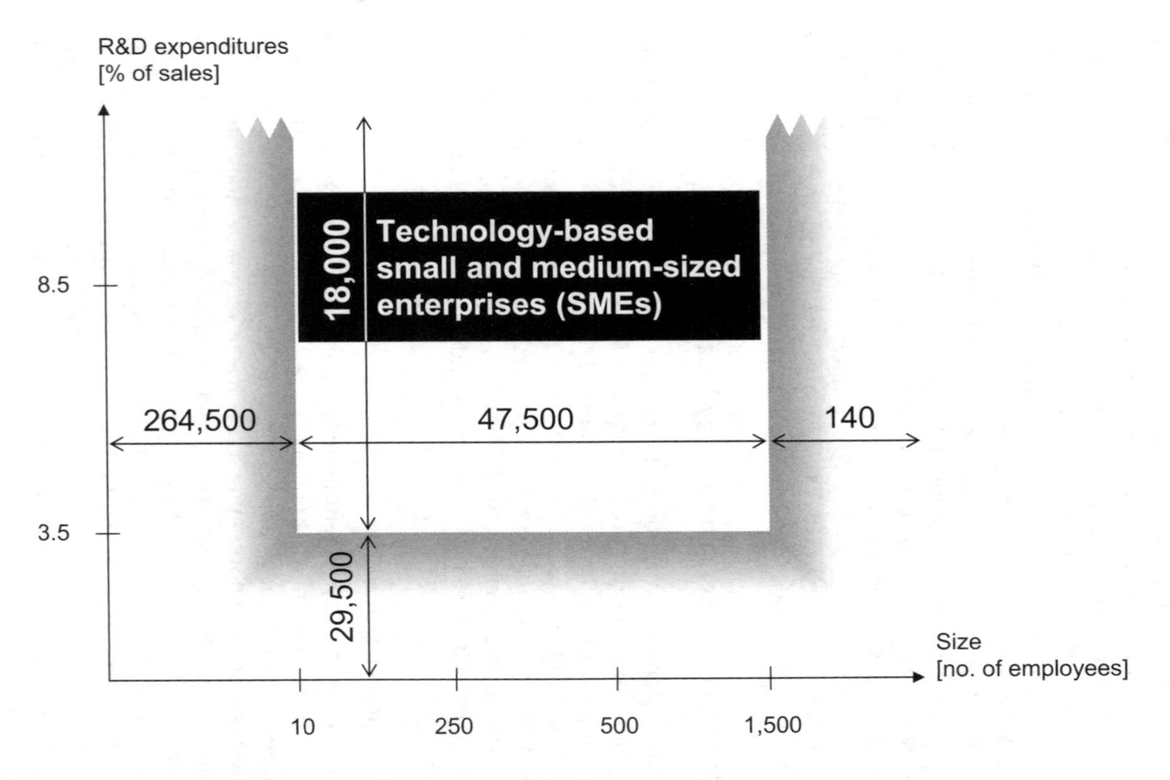

Figure 2.7 Estimation of the number of technology-based SMEs in Switzerland

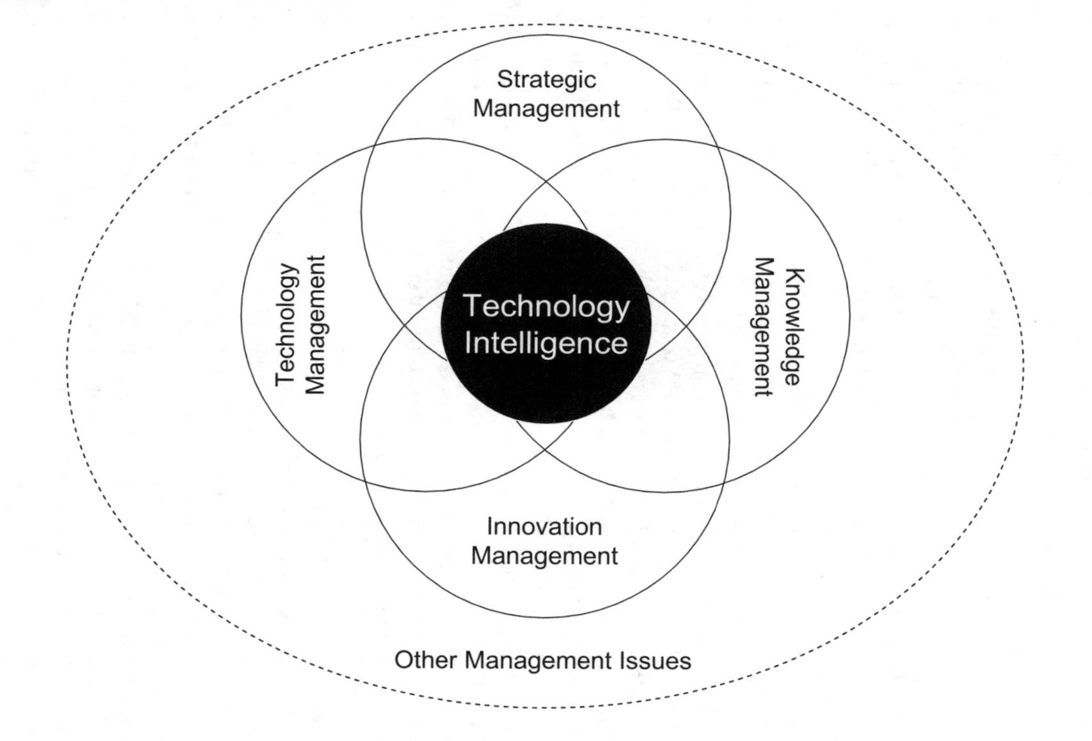

Figure 2.8 Management issues relevant to Technology Intelligence

- **Design school**: A process of conception. Strategy formation achieves the essential fit between internal strengths and weaknesses and external threats and opportunities. Senior management formulates clear, simple, and unique strategies in a deliberate process of conscious thought. This process is neither formally analytical nor informally intuitive.
- **Planning school**: A formal process. This school of thought is close to the design school with a rather significant exception: the process is not just cerebral but formal and can be broken down into distinct steps. It is delineated by checklists and supported by techniques. This means that staff planners replace senior managers, *de facto*, as the key players in the process.
- **Positioning school**: An analytical process. Strategy reduces to generic positions selected through formalized analyses of industry situations. Hence, the planners become analysts.
- **Entrepreneurial school**: A visionary process. Much like the design school, the entrepreneurial school centers the process on the chief executive; but unlike the design school and in contrast to the planning school, it roots that process in the mysteries of intuition. This shifts strategies from precise designs, plans, or positions to vague visions or broad perspectives.
- **Cognitive school**: A mental process. Strategies originate from mental processes of cognition by information processing, knowledge-structure mapping, and concept attainment. Thus, cognition is used to construct strategy as creative interpretations, rather than simply to map reality in some more or less objective way, however distorted.
- **Learning school**: An emergent process. The model of strategy-making as learning is based on ideas such as disjointed incrementalism, logical incrementalism, venturing, emergent strategies, and retrospective sense-making. Strategies are emergent, strategists can be found throughout the organization, and so-called formulation and implementation intertwine.
- **Power school**: A process of negotiation. Strategy making is rooted in power at two levels: micropower sees the development of strategies within the organization, macropower views the organization as an entity that uses its power over others and among partners in alliances, and vice versa.
- **Cultural school**: A social process. In parallel to the power school, which focuses on self-interest and fragmentation, the cultural school focuses on common interest and integration. Thus, strategy

formation is a social process rooted in culture. This school is particularly attributed to Japanese organizations.
- **Environmental school**: A reactive process. Not strictly a strategic management school, but always of strategic management concern, is illuminating the demands of environment. For example, the contingency theory considers which responses are expected of organizations facing particular environmental conditions that claim limits to strategic choice.
- **Configuration school**: A process of transformation. This school sees organizations as configurations – coherent clusters of characteristics and behaviors – and integrates the claims of the other schools: each configuration in its own place. Consequently organizations can be described by states, for example as a mature or start-up company, and change must be described as transformation that brings the organization from one state to another.

Across these 10 schools of strategy formation, various approaches become a hybrid by linking some or all of the elements of the different schools (Mintzberg & Lampel 1999: 26). Examples are: stakeholder analysis (linking the planning and positioning schools), chaos theory (being a hybrid of the learning and environmental schools), dynamic capabilities (being a hybrid of the learning and design schools), and resource-based theory (being a hybrid of the learning and cultural schools).

The various schools and approaches having been established, the emerging question is if they represent different strategy processes or complementary parts of the same process. Mintzberg & Lampel (1999: 27) comes to the conclusion that most schools partly represent aspects of what can be referred to as strategy formation, which is in sum "judgmental designing, intuitive visioning and emergent learning; it is about transformation as well as perpetuation; it must involve individual cognition and social interaction, cooperative as well as conflictual; it has to include analyzing before and programming after as well as negotiating during; and all this must be in response to what may be a demanding environment" (ibid.: 27).

Abell (1999) presents an alternative view, and makes strategy formulation more concrete by suggesting consideration of dual strategies which are run in parallel: "today-for-today strategies" and "today-for-tomorrow strategies."

This distinction between a present and future orientation is not the usual short-term, long-term distinction – in which the short-term plan is simply a detailed operations and budgeting exercise made in the context of a hoped-for long-term market position. Present planning also requires strategy – a vision of how the firm has to operate now (given its competencies and target markets) and what the role of each key function will be. The long-term plan, in contrast, is built on a vision of the future – even more important, on a strategy for getting there. (Abell 1999: 74)

Of course there is some interaction between these two strategies. In particular, today-for-tomorrow strategies lay the foundation for today-for-today strategies. Even if the emphasis is on designing or planning the strategy, other strategy formulation schools are complementary, or even necessary, in such a dual consideration.

Because strategic management is obviously a balancing act between internal competencies and external (market and technology) demands, Technology Intelligence is a concern in strategic management. This interaction and interdependency will be explored in detail later on, when the mission and goals of Technology Intelligence are discussed.

Much of the literature on the role of strategic management in SMEs is ambiguous. While some writers argue that "formal strategic management procedures are inappropriate for SMEs which have neither the management nor financial resources to indulge in elaborate strategic management techniques, and for companies operating within the turbulent environment of high technology industries where conditions change so fast that environmental forecasting is of questionable value" (Berry & Taggart 1998: 887), empirical work in successful, rapid-growth, technology-based SMEs concludes that in those firms, strategy formulation is important, if not essential (Dodgson & Rothwell 1991; Berry 1996). However, there are two tendencies upon which the literature agrees: first, the implementation of an effective strategy development process depends heavily upon the characteristics of the entrepreneur (Gibb & Davies 1990: 29); and second, the more the firm grows (or the larger the firm is), the more strategy activities become formalized (Berry & Taggart 1998: 893).

This short literature review shows that next to the obvious assignment of the entrepreneurial (and probably power) school, any of the other schools are of interest in the SME context. Hence, Mintzberg and Lampel's (1999) and Abell's (1999) observations, discussed above, are acceptable for technology-based SMEs.

Technology management

The purpose of **technology management** is the deliberate handling of technologies. For decades, several authors have been developing various approaches to technology management. This book is inspired by Tschirky's (1998b: 267) concept of "integrated technology management" (ITM), an approach motivated in turn by Bleicher's (1992: 223) concept of "integrated management," adapted for the technology dimension of companies. The vision of the concept of ITM is "Bringing Technology into Management." "Its basis is the postulate that 'technology issues' will no longer be solely of concern in the context of direct technology-related managerial functions such as R&D and production management but will be of prime concern for general management at all levels" (Tschirky 2000: 417).

Tschirky (1998b: 269) distinguishes between three levels of management:

- On the *normative level*, a clear commitment to the importance of technology as a vital concern of the technology-based company should be anchored in the company policy. At the same time, technology awareness should permeate company culture at all levels.
- The *strategic level* is the transformation of company policy to comprehensive (technology) strategies. The principle of effectiveness is dominant. A primary concern is the trilogy of technology decisions: (a) Which way to go?, (b) Make or buy?, (c) Keep or sell? Of course other management aspects, like marketing and resource procurement, strongly relate to these decisions.
- Finally, on the *operational level*, responsibility is taken for transforming strategies into practice in the context of short-term goals. Milestones in R&D projects, resource allocation, and formal information flow are, for example, typical tasks at this level. The principle of efficiency is essential.

According to this view, technology management can be conceived of as an integrating function of general management which is directed towards the normative, strategic, and operational management of the technology and innovation potential of an organization (see Figure 2.9).

Technology intelligence is typically of concern at the strategic and the operational levels of technology management. This is particularly true for the lateral integration of technological and societal change as well as the company development.

The importance of TI to technology management is demonstrated in several studies. In Gupta and Wilemon's (1996: 507) study about key knowledge domains in industrial R&D management, monitoring science and technology developments is ranked among the top ten domains. Scott's findings (2000: 64) are comparable. In his Delphi study, technology trends and paradigm shifts are also ranked in the top ten. The top-ranked issue was strategic planning for technology products. A follow-up Delphi study was completed to clarify the nature and dimensions of the top problem. In this second study, TI activities and technology knowledge acquisition were ranked as very important issues in strategic planning for technology products (Scott 2001: 16).

In general, Technology Intelligence can be understood as an information delivering task, and thus one which supports technology management.

Kohler (1994: 216) examines the technology management situation in Swiss SMEs and offers the following important insights:

- Technology management goals are not well developed in SMEs.
- There is no correlation between technology management occurrence and company success in SMEs.
- The technology management presence is stronger in more innovative SMEs than in less innovative ones, especially in medium-sized companies.
- The technology management presence is stronger in medium-sized companies than in small ones.

Technology management is not prevalent in SMEs. In this study it will be shown that companies that care about technology management, i.e. companies who deliberately handle technology issues, are more innovative than others. However, a direct benefit is not proven. Other studies explicitly examining technology management in SMEs are very rare, although some studies on other management issues (such as innovation management) in SMEs cover some aspects of technology management. Some contributions are discussed in Kohler (1994: 147) and in the next chapter.

Innovation management

The term "innovation" has become very popular. Mitterdorfer (2001: 13) gives a chronological overview of the term and its significance (see Table 2.3). The term has evolved from a focused to a widely used, success-oriented term.

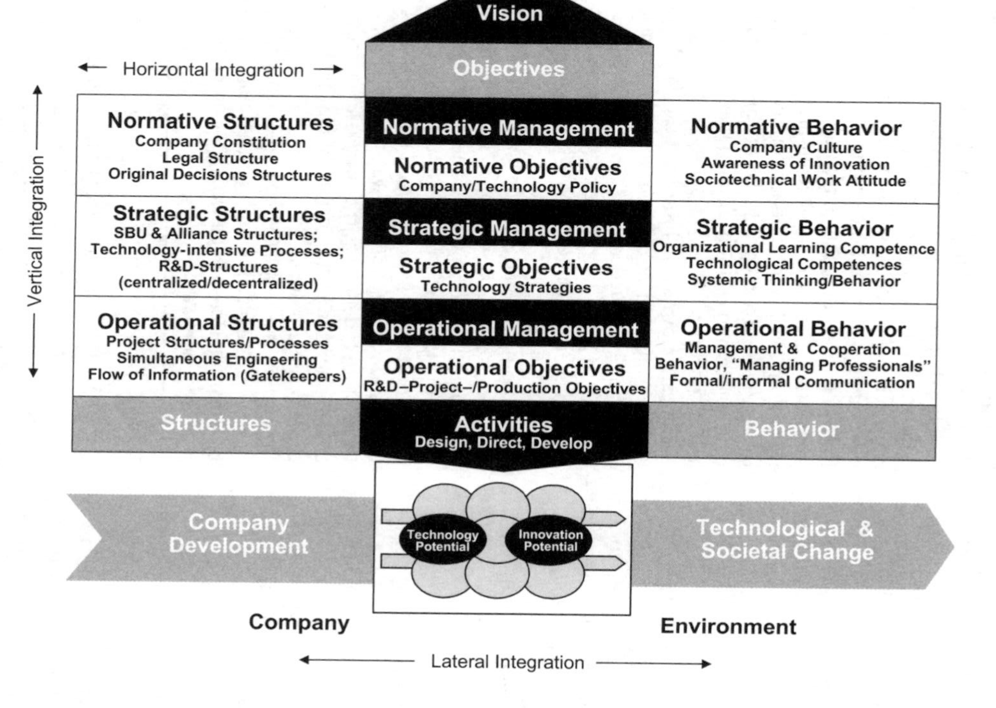

Figure 2.9 The concept of "integrated technology management" (Tschirky 1998b: 270)

Table 2.3 A chronological overview of the term innovation

Author	*Description*
Schumpeter (1939: 87)	If, instead of quantities of factors, we vary the form of the production function, then we have an innovation.
Barnett (1953: 7)	An innovation is defined as any thought, behavior, or thing that is new because it is qualitatively different from existing forms.
Schmookler (1966: 2)	An enterprise produces goods or services or uses a method of input that is new to it.
Marquis (1969: 1)	Innovations are the units of technological change.
Witte (1973: 17)	Innovation is the first-time (economic) use of an invention. The invention does not necessary have to emerge from R&D or science, but also includes new objects and methods from management and social science.
Pavitt (1980: 1)	Technical innovation in industry is the development, commercialization, adaptation, and improvement of product and production processes.
Drucker (1985: 67)	Innovation is the effort to create purposeful, focused change in an enterprise's economic or social potential.
Foster (1986: 20)	Innovation is the battle in the marketplace between innovators or attackers trying to make money by changing the order of things, and defenders protecting their existing cash flow.
Ergas (1987: 191)	Innovation is the use of human, technical, and financial resources to find a way to do things.
Udwadia (1990: 66)	Innovation is defined as the successful creation, development and introduction of new products, processes, or services.
Vahs & Burmester (1999: 1)	Innovation is a goal-oriented implementation of new technical, economic, organizational, and social problem solutions which leads the company to its goals in a new way.

Innovations are subject to an innovation process. Literature provides numerous varying views on the innovation process, how it is organized and structured in phases, stages or steps. Common to all of them is that at the beginning there is something like an idea, and at the end a kind of realization and commercialization of the idea. In the present book, the approach of Savioz et al. (2002) is accepted (Figure 2.10). With regard to TI, especially the early stages of innovations, the so-called "fuzzy front-end" is of primary concern because TI might give an impulse and an input to idea generation. However, Technology Intelligence supports the innovation process at all times.

Innovation management deals with the organization or coordination of the innovation process (Hauschildt 1993: 25). The innovation process is not necessarily formal. In the case of SMEs the process is often realized informally (Minder 2001: 82). A large number of contributions on innovation management in SMEs exist. Among others, the most promising advantages of SMEs with regard to large companies are favorable company culture and high flexibility. In contrast, insufficient resources of any kind limit innovation opportunities. But in general, the knowledge base with regard to how SMEs actually undertake innovative activities remains very limited (Hoffman et al. 1998: 39).

Knowledge management

The definitions of **knowledge management** range from narrow, utilitarian views to much more conceptual and broad perceptions (Raisinghani 2000: 107). Minder (2001: 61) discusses three definitions:

- Textual definitions: the goal of knowledge management is to create and use knowledge in order to increase and maintain added value (Schneider 1996: 41). Chrobok (1998: 184) sees the task of knowledge management as selecting, collecting, analyzing, summarizing, and using internal and external knowledge to serve the organization in any area. Gomez's (1996: 1) definition is broader: knowledge management shows how to process and use corporate knowledge in an optimal way.
- Knowledge management as continuation of organizational learning: knowledge management tries to give management instruments of intervention in the corporate knowledge base by means of concepts and tools. Of particular interest are the parts of the learning process which are manageable (Probst et al. 1999: 46).
- Operational and technical aspects: Becker (1995: 16) defines knowledge management as designing business processes and building information and communication technology (ICT) infrastructure as an environment within which knowledge develops.

However, a common thread seems to be apparent: knowledge management is commonly defined as an attempt to put processes in

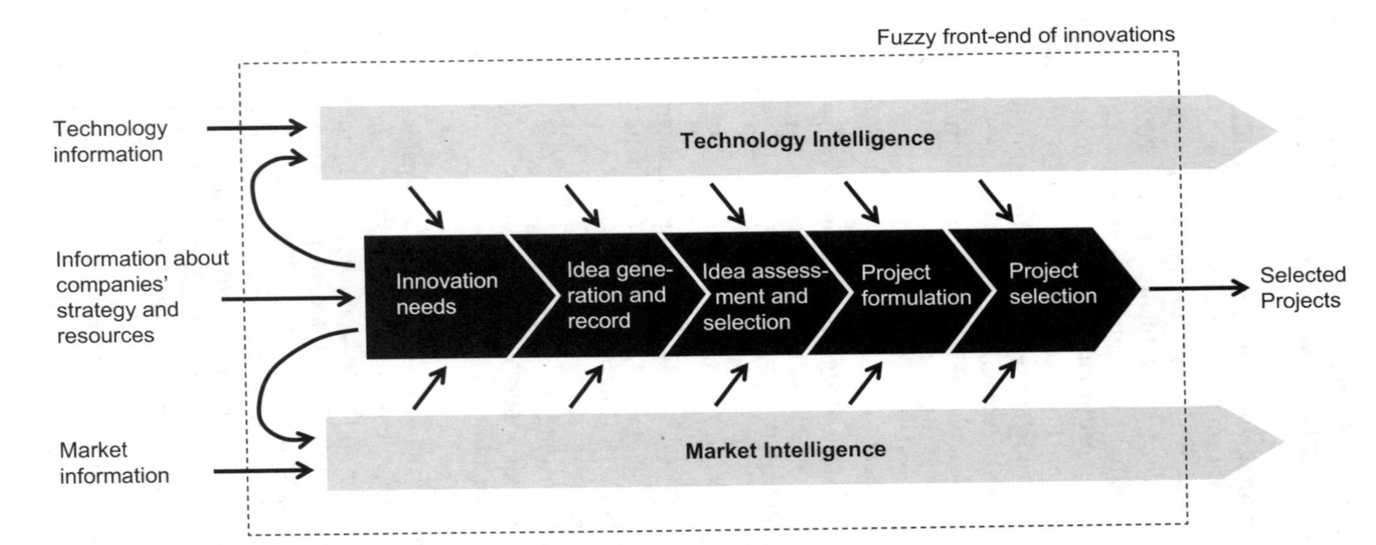

Figure 2.10 The "fuzzy front-end" of innovations
Sample of the TopNano21 project

place that capture and reuse an organization's knowledge base so it can be further utilized to generate revenue (Raisinghani 2000: 107). The present study adopts this definition.

It is surprising that in spite of the exploding amount of literature on knowledge management, little is dedicated specifically to SMEs.

However, managing the value that can be derived from knowledge is a challenge for all businesses, large and small. Understanding of organizational theory and practical considerations of knowledge management has largely been derived from large company developments. It seems clear that small and medium-sized businesses will need to address their knowledge management practices, but that, like in so many aspects of business and management, the issues that small and medium-sized businesses will face will not be simply a scaled-down replica of large-company experiences. (Sparrow 2001: 3)

Sparrow (2001) discusses features of small firms which when combined constitute a different milieu for knowledge management. The case-study work has highlighted the fundamental need to recognize the mental models that participants have, and the need to work towards means of eliciting and sharing personal understanding. In small firms especially, knowledge systems need to amplify human potential. Thus, knowledge management requires an appreciation of individual and collective understanding. Sparrow sees the adaptability of small firms as an advantage: The articulation of the knowledge base through the development of a formal organization system for knowledge identification, definition and evaluation, as an approach to knowledge management, is rooted in the philosophy of predictability, measurement and control. An alternative philosophy might place less emphasis on the de-contextualization of knowledge through its description and measurement, and more upon chaos theory notions of self-correcting, flexible and dynamic knowledge processes through which the business secures its adaptability. There is evidence to suggest that this is a characteristic of management approaches in small firms. (Sparrow 2001: 7)

Furthermore, Sparrow proposes a model for the development of knowledge-management practices in SMEs. The model counts different foci which vary with the phase and the stage of the knowledge project. However, the Sparrow's model remains descriptive.

Minder (2001) conducted another study in SMEs, where she also concluded (2001: 348) that knowledge-management initiatives should always be adapted to their specific needs. For example, what happens if an experienced researcher leaves the company? This is particularly harmful to SMEs because of their limited (knowledge) resources. This in turn requires adaptive measures.

3 The Practitioners' Voice

The introductory chapter set the focus of this book on theoretical argumentation. The aim of this chapter is to clarify the relevance of the topic in practice. It is not our goal here to examine practices in companies, but to examine interest in the topic. The practitioners' voice is captured by means of interviews, and through seminars and workshops. Since the first are specific to companies and the others are group events, they are discussed separately.

Interviews

Several interviews in Swiss companies were held in the context of the TopNano21 project.[1] The sample was heterogeneous, ranging in size from 15 to 7,000 employees, and from less than 2 percent R&D expenditure to more than 15 percent. About 10 companies coped with the definition of technology-based SMEs (Figure 3.1). Henceforward, we

---

1 The Swiss government recognized the need for fundamental research in the area of nanotechnology and its management, and in 1998 launched a program called TopNano21, which coordinates and finances some 200 industrial and academic research projects dealing with nanotechnology. One of them is the "Nanotechnology Innovations Check: A Tool for Identifying and Evaluating Enterprise Specific Application Fields within Nanotechnology" (KTI Project 5049.1). The aim of this project is to develop an instrument allowing early recognition of the possible use of nanotechnology in a company's own products and processes, to reliably assess the opportunities and risks of possible activities in the area of nanotechnology, to derive company-specific (nano)technology strategies, and to integrate scientific and technological knowledge into early stages of the innovation process. The interviews' focus was on the impulse to, and identification, observation, assessment, selection, and implementation of, nanotechnology.

will be considering these companies in our discussion. The companies to be examined are well distributed in the field of interest. Most of them (Actelion, Komax, Mathys Medical, Nägeli, Phonak, Straumann, SwissOptic, and Zeptosens) are independent; some (Novartis Opthalmics and Roche Diagnostics Instrument Center) are affiliated with large multinational companies. However, the latter also operate rather independently from the holding company – i.e. as profit centers.

A major insight gained from these interviews was that most companies are facing the problem of getting relevant information from outside the company. Some companies pursue certain "intelligence" activities, but they are in general not consciously doing so. Most companies complain about difficulties in diverse stages of the value-creation process of Technology Intelligence (TI) (see Chapter 4). To summarize the most important problems and needs:

- *Formulation of information needs*: Availability of information about technological and market facts and trends seems to be crucial for a company's success. However, companies rarely explicitly formulate their information needs. Companies have difficulty in seizing the innovation potential of trends. On the one hand this is because companies gather information about existing customers, but not about potential customers. Thus they are often obliged to be reactive with regard to any major change. On the other hand, companies face problems estimating the problem-solving potential of technological changes in their own business. To summarize, there is a need to increase the awareness of what information is useful to a company, and to shed light on information needs in general.
- *Information collection*: Most companies confirmed that they use formal as well as informal information sources. Most companies subscribed to some scientific journals and databases, and they attend conferences and seminars. Normally, they also entertained an informal network of experts. However, there is no insight into how intensely these sources were used. In general, both researchers and managers gather information continuously in order to keep technological evolution on track. If any technology is of particular interest, most companies generate a project in order to get specific information about the technology. Sometimes they refer to external experts. A major problem seems to be the transfer of knowledge between different people, especially between researchers and managers. Thus, information is available, but is poorly matched with strategy. Another difficulty, especially in the case of management

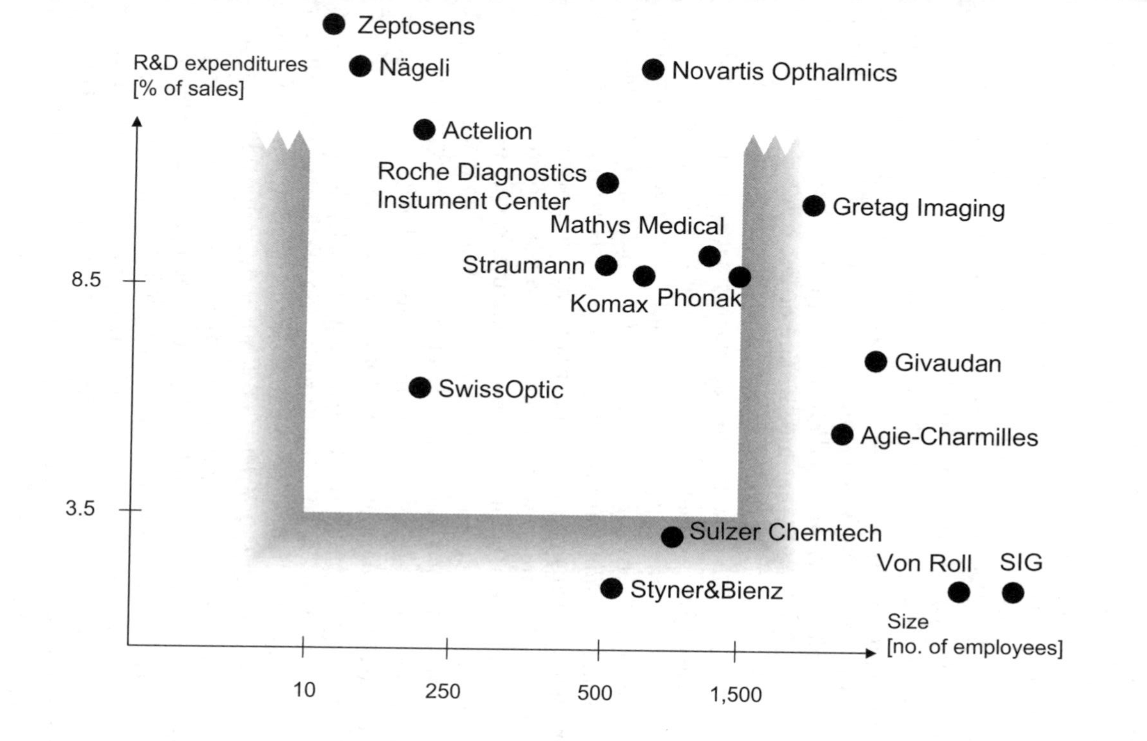

Figure 3.1 Sample of the TopNano21 project

involvement, is the lack of time for information collection. To summarize, information is generally accessible, but is not necessarily considered for decision-making.

- *Information analysis*: The interviews showed that there is a strong need to improve the analysis stage. Though analysis is commonly done in an interdisciplinary manner and at all company levels, the use of analysis tools is poor. There is a call for standardized methods. In general, analysis occurs step by step, which means that only relevant information continues to be analyzed. Together with several (resource-reducing) advantages, some potentially relevant information might get lost. To summarize, information analysis remains unsatisfactory in most companies, especially due to the lack of tools.
- *Information dissemination*: Most companies have a kind of "Technology Board" that meets more or less regularly (weekly to twice a year). Thus, facts and trends that are relevant to the company to any extent can be disseminated in this fashion. Normally the information reaches business strategy because top management is mostly part of the "board." However, such meetings are insufficient by far for effective dissemination of information. Informal communication, which comes along with an open and modern company culture, is important. There were some differences between the interviewed companies. While some are downright famous for their favorable company culture, others seem to have problems with communication barriers. Most companies would welcome a defined communication logic, but they always insist on the importance of an informal information exchange.
- *Information application*: Since normally there is no systematic Technology Intelligence in these companies, there is a barely defined TI client. In addition, information seems to be more or less well disseminated. Therefore, anyone can apply the information. This corresponds to a modern view of "organizational intelligence" and should be encouraged. In fact, the application of information begins early – for example, to decide whether a trend should be pursued or not. However, no insight could be gained about the extent to which the information is used, and what value it has. To summarize, there is little insight into how information is applied, though most companies agree with the fact that decisions should be based on solid information.

The discussion above has demonstrated that companies are aware of the problem of gathering relevant information about their technological environment. Furthermore, they take other forms of action, which suggests that Technology Intelligence is not a "virtual" but a "real" concern. There are some differences in TI practice between the considered companies. However, there was no detectable pattern for TI system design or implementation.

Companies commonly are less than completely satisfied with their current situation. They are often uncertain when making a decision. Therefore, there is a call for more systematic Technology Intelligence. This does not necessarily mean more formal activities, but rather giving the activities a logic through finding a way to coordinate them.

Seminars and workshops

Some seminars and workshops where Technology Intelligence was the main topic or part of a more general topic (for example, technology management) were organized at the Swiss Federal Institute of Technology, at the ETH Center for Enterprise Sciences in Zurich. Since some seminars had already presented certain conceptual approaches, among them approaches developed later on in this book, the audience was biased for discussion. However, since the goal of this chapter is to get the practitioner's reality, two points are of interest. First, the brisk interest of participants demonstrates the urgent need for Technology Intelligence in practice, and also demonstrated the lack of "ready-to-use" concepts. Secondly, the general insight, problems, and needs of practitioners can and will be pointed out. In addition to the ETH seminars, another seminar on "competitive intelligence" at Centredoc gave some insight into TI awareness in Swiss SMEs. Centredoc is a information broker specializing in the watch industry. The company is a partner of FSRM – the Swiss Foundation for Research in Microtechnology.

The Swissmem–ETH community of practice: technology management

This community of practice in technology management is a partnership between the Swissmem and the ETH Center for Enterprise Sciences, Technology, and Innovation Management which has existed since 1996. ("Swissmem" stands for Swiss Mechanical and Electrical Engineering Industries, a union of two associations – the ASM, or Association of Swiss Engineering Employers, and the VSM, or Swiss Association of Machinery Manufacturers – which represent to politicians, national and international organizations, employee groups, and the public, the interests of their nearly 1,000 member companies. As the sector's foremost representative, Swissmem offers practical services such as advice on export matters, support on employment law issues, sector-specific basic and ongoing training programs, and activities tailored to the sector's division.) There are two meetings a year to exchange theoretical insights and discuss practical needs. Two sessions relating to Technology Intelligence were held at ETH in Zurich. The first, on April 27, 1999, concerned "Concepts of Technology Intelligence." There were 6 participants from 5 companies (ABB Consulting, Ascom, Hilti, Huber&Suhner, VSM). The speakers were Prof. Dr. Hugo Tschirky and Dr. Eckhard Lichtenthaler. The aim of the session was to sensitize the audience to the importance of being aware of their technological environment and changes therein, and to demonstrate TI solutions for large multinational companies (i.e. Daimler-Chrysler). The discussion led to the following conclusions:

- The companies had, and were aware of, problems gathering relevant information about their technological environment.
- The companies were already engaging in activities in this field, but they were not coordinated. The participants wanted to approach the problem more systematically.
- A "large" solution is not practical for most companies because resources are not available (even for large companies – for example Hilti, with 12,000 employees).

The second session on March 8, 2001, was about "Technology Management in SMEs." There were 8 participants from 8 companies (Alu Menziken, Ammann, Ascom, Contraves Space, Siemens Switzerland, SIG, Styner&Bienz, Von Roll). The speakers were Hugo Tschirky, Beat Birkenmeier, Dr. Harald Brodbeck (B'Results Ltd.), and Dr. Pascal Savioz. The aim of this session was to emphasize the importance of technology management to SMEs, and to show the audience two applications in industry: structuring of the fuzzy front-end of innovations and Technology Intelligence in two medium-sized companies. Lessons learned from discussion, most notably from the TI part, were:

- As in the previous session, companies did in fact engage in some TI activities, but they were in general not aware of doing so. There was also a wish to systematize these activities to a greater extent.
- Top management should definitely be involved in TI activities.

- The most important problem seemed to be information flow. There was little knowledge about which (internal and external) information was available, and where one could find this information. It was in general not clearly known to whom the gathered information should be directed.
- An approach to handling complexity seemed to be a steady cycle of reduction and increase of information quantity.

The seminar on "Technology Intelligence in SMEs"

A seminar on TI in SMEs in the series "Technology, Innovation, and Management" was held on the April 6, 2000, in Zurich. There were 10 participants from 8 companies (Afag, Alaxa Consulting, Alveo, Ammann, Baer, Hesco, Sefar, UBS). The speaker was Dr. Pascal Savioz. The aim of the seminar was to give a general introduction to the topic of Technology Intelligence, to illustrate solutions for TI systems in large companies, and to point out some challenges for SMEs. Lessons learned from this discussion were:

- Most companies did not have any experience with intelligence activities. Certainly they sought this information, but they were not aware of value creation when information is well used.
- They agreed with the importance of coping with technological change by means of observation of their technological environment, and they were convinced of the effectiveness of a systemic approach like a TI system.
- However, a major problem seemed to be resources. First of all, companies saw a lack of managerial competencies for setting up and maintaining such a system.

The seminar on "Competitive Intelligence" at Centredoc

The seminar was held in four sessions on May 2, June 27, September 26, and October 24, 2000 at Centredoc in Neuchâtel. There were 18 participants from 14 companies (Aton Technologies, Creapole, Esnig, Etel, FKG Dentaire, ID Informatique & Developpement, Imprimerie Corbaz, Innosyn, Lem, Patek Philippe, Plastiglas, Reymond, Sonceboz, Vibro-Meter). The speakers were employees of Centredoc. The subject was how to set up a "competitive intelligence cycle." There was a strong emphasis on information sources and software (i.e. v-strat) that supports the cycle. The seminar was not interactive, thus it was hard to deduce lessons from the practitioner's reality. However, the number of participants gives some evidence of interest in Technology Intelligence among SMEs.

Conclusion

The aim of this chapter was to look at insights gained from practitioners about the problems and needs of SMEs in relation to Technology Intelligence. Indeed, TI is of significant concern in SMEs' reality, as several interviews and the numerous participants in diverse seminars and workshops (in total 35 companies) showed. Apparently there are no existing concepts for gathering and processing relevant information which cope with SME-specific needs. However, there is not a complete lack of TI activities in SMEs. Therefore the **conclusion** is:

> *Some SMEs already pursue (implicitly or explicitly) Technology Intelligence activities, but these efforts are rarely coordinated. However, most companies wish to give the activities more structure, for example with a Technology Intelligence system: a call from reality!*

There is a general skepticism among SMEs about the transfer of "big" solutions designed for large firms. In reality, these solutions seem to be too resource intensive, considering both knowledge and capital. Thus, from a practical point of view, there is a need to design a Technology Intelligence system which meets SMEs' specifications.

4
Technology Intelligence – An Overview

Business intelligence and competitive intelligence

The terms "business intelligence" (BI) and "competitive intelligence" (CI) are often used interchangeably. Both BI and CI refer to "actionable information about the external business environment that could affect a company's competitive position" (Ashton & Klavans 1997: 9). A similar overlap between Competitor and Competitive Intelligence, and between Business Intelligence and Environmental Scanning, is observed by Choo (1998: 81). He argues that since competitive intelligence is a primary objective of activities within business intelligence, those two terms can be used more or less interchangeably. Figure 4.1 attempts to depict the relationship between the different terms.

Competitor intelligence is information-gathering about the actual and future activities of competitors, whereas **competitive intelligence**'s focus is broader and embraces Porter's (1980) five competitive forces model. **Business intelligence** is "the activity of monitoring the environment external to the firm for information that is relevant for the decision-making process in the company" (Gilad & Gilad 1988: 14). Thus, BI is concerned with exploring possible views of future competitive environments. This is almost the same scope as **environmental scanning**, which is "the acquisition and use of information about events, trends, and relationships in an organization's external environment, the knowledge of which would assist management in planning the organization's future course of action" (Auster & Choo 1994: 607). According to Ansoff (1980), **issue management** tries to identify future trends in time to prevent a crisis. "A strategic issue management system is a systematic procedure for early identification and quick response to important trends and events both inside and outside an

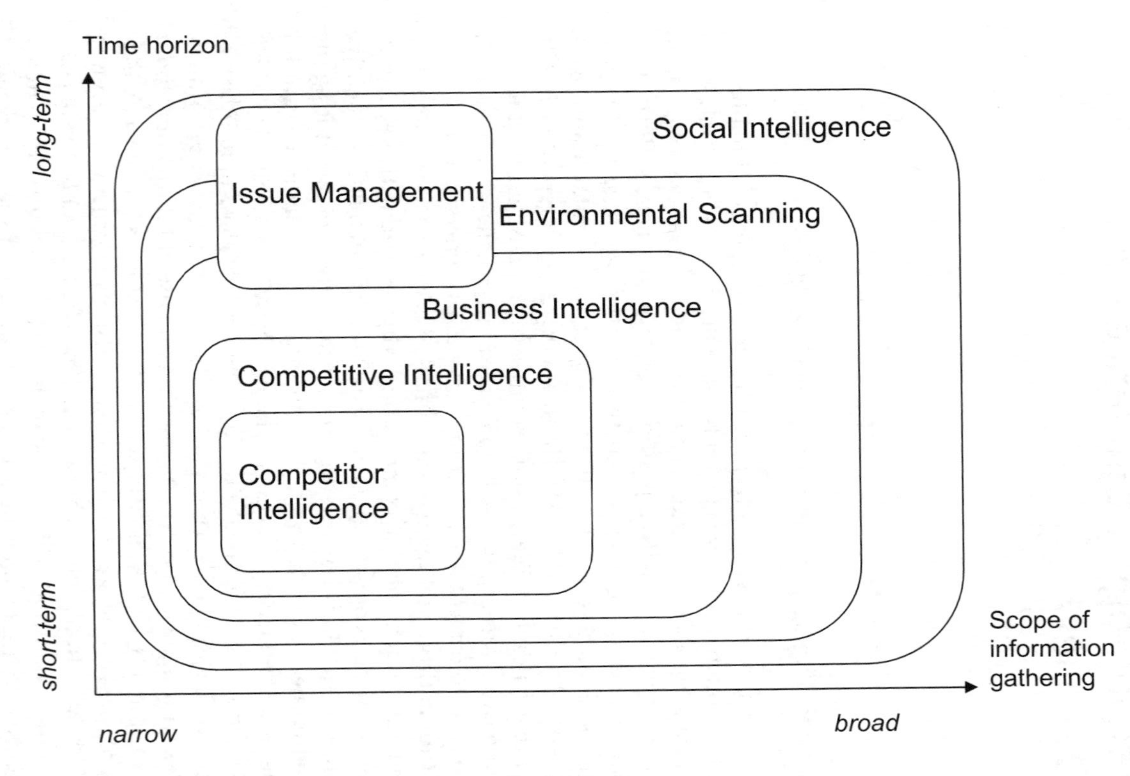

Figure 4.1 Forms of organizational external information-gathering

enterprise." Finally, **social intelligence** is the broadest in scope and approach, and is concerned with the capability of society and institutions to identify problems, collect relevant information about those problems, and transmit, process, and evaluate, as well as ultimately put this information to use (Dedijer & Jéquier 1987: 34).

This discussion shows that the borders of these domains are rather blurred. It would be confusing to deal with all these definitions. Therefore, for the purposes of this book, business intelligence will be defined as follows:

> *Business intelligence consists of activities of collection, analysis, and application of information describing relevant facts and trends (opportunities and threats) from the organization's entire environment used to support the business decision-making process.*

Is business intelligence a part of strategic planning, or is it a parallel process? The interaction between these two processes is certainly intense. Strategic planning determines the focus of observation, whereas BI brings insight into planning alternatives. In a way, one could describe it as a "chicken–egg" phenomenon. In this book we accept both views: while BI is a part of strategic planning from an inside-out point of view, it is a parallel process from an outside-in point of view. Furthermore, this implies that identification of relevant trends need not necessarily be early but can simply be timely.

Technology intelligence – underlying concepts and definitions

In this book, **Technology Intelligence** (TI) is treated as a part of business intelligence. However, as in the case of BI, in the literature on the subject there is a controversy about the meaning of the term and about what TI encompasses.

This chapter will clarify how the term "Technology Intelligence" is used in the present book, and illustrate some aspects of the usage as well as its underlying concepts. The chapter concludes with a working definition of TI.

Technology monitoring

Technology monitoring deals with specific or unspecific observation of the (technological) environment for pertinent information (Porter et al. 1991: 114). This involves "identifying signals of change in embryonic stages, and gathering information on appropriate phenomena and parameters to determine the rate of advancement as well as the character and form that potential impacts of the change might take" (Utterback & Brown 1972: 6). Thus, technology monitoring acts as an alarm system of potential changes, which means opportunities or threats. The challenge is to handle the observation area in scope and time: observing too narrow an area limits the probability of detecting opportunities, while it is dangerous to be unaware of threats from outside these limits. In comparison, observation that is too general is very resource-intensive. To determine whether an observation is an opportunity or a threat, this information has to be analyzed and assessed, which leads us to technology forecasting.

Technology forecasting

Technology forecasting (TF) goes beyond observation of signals and events; it assesses these signals and events in accordance with business strategy (Twiss 1992: 53). The literature on forecasting (i.e. assessment) methods is voluminous. A common critique of these methods is that the output of these mostly mathematical models is on (data-)products rather than on strategy-related intelligence. Therefore, the scope of modern technology forecasting has gradually been extended. Modern TF is viewed as a learning process, which analyzes and communicates the TF results.

Some authors use the following definition for TF: "Technology Forecasting deals with causal elements of any sort – social, economic, or technological. However, the effects of interest are new technologies, and incremental and/or discontinuous changes in existing technologies. The focus of the analysis is on a technology or a family of technologies. Thus the end processes are the development of technological principles and/or prototypes and the diffusion of technological devices" (Porter et al. 1991: 21). This underscores the importance that the observation area go beyond technological issues.

Technology scouting

Technology scouting deals with "seeking new ideas in science and technology for further development" (Brenner 1996: 20). Unlike technology monitoring and forecasting, technology scouting collects and screens targeted information on particular technologies, experts, or organizations in response to the user's request. This is particularly important in the case of the increasing external technology acquisition.

Scouting is an operational rather than a strategic task, and therefore differs from monitoring and forecasting. However, it is complementary to these activities for two reasons: first, the process of scouting resembles the others; secondly, a proactive search makes new insight about trends possible. Thus, scouting might be accepted as a part of TI.

Competitive technical intelligence

Competitive technical intelligence (CTI) is a practice-oriented approach, which includes both the process and the result of information-gathering and analysis. Thus different definitions of CTI sometimes are used even by same author: "Competitive Technical Intelligence is the practice of collecting, analyzing and communicating the best available information on science and technology developments and trends occurring outside one's own company" (Ashton & Stacey 1995: 82) and "Competitive Technical Intelligence is business-sensitive information on external scientific or technological threats, opportunities, or developments that have the potential to affect a company's competitive situation" (Ashton & Klavans 1997: 11). The main objectives of CTI are:

- To provide early warning of external technical developments or company moves that represent potential business threats or opportunities;
- To evaluate new product, process, or collaboration prospects created by external science and technology activities in time to permit appropriate responses;
- To anticipate and understand science- and technology-related shifts or trends in the competitive environment as preparation for organizational planning and strategy development.

Discussion of underlying technology-intelligence concepts

The boundaries of the research fields discussed above are quite blurred. Depending on their school of thought, authors tend to define the core element of their research as dealing with "getting information to strengthen the organization's planning for the future," and then extend that definition with missing elements. This is true with regard to both terminology and content (goals, activities, etc.).

However, there are two main schools of thought. On the one hand, there are authors who present methods with the aim of predicting technological development in the future. On the other hand, there are attempts to develop systems which periodically or occasionally observe an organization's technological environment in order to assess the impact on it. These two approaches are complementary, but how?

It would be dangerous to neglect the insights gained in each of these fields after years of detailed research. Nevertheless, once again, we need to declare a definition of TI for the purposes of this book.

Definition of Technology Intelligence

The definition of TI in the present book will be as follows:

> *TI activities are those activities which support decision-making of technological and general management concerns by taking advantage of a timely preparation of relevant information on technological facts and trends (opportunities and threats) of the organization's environment by means of collection, analysis and dissemination.*

Here are some parts of the definition which deserve highlighting:

- *TI supports decision-making in technological and general management concerns*: Decision-making not only depends on the most reliable information, but also on intuition, traditions, resources, etc. Thus, TI is a supporting task, which can be pursued systematically or informally. Since decisions can be made within a planning process or spontaneously, TI has a reactive as well as proactive character. Technological trends may have an impact on any potential of the organization. Therefore, TI influences both technological and general management concerns.
- *Technological facts and trends (opportunities and threats)*: There is no restriction on the time focus or on the observation area. While facts express a status, trends describe the evolution of this status. This fundamental differentiation seems to be missing in recent intelligence research. These facts and trends may emerge from weak signals and have a long-term scope. In turn, there may be established technologies to solve a current problem. Thus, TI deals with strategic and operational questions. Whether facts and trends are opportunities or threats depends on their perception. In any case, an integrated view comprising both technology and the market should be considered.
- *Well-timed preparation of relevant information*: Today's information might have another significance tomorrow because the organization's context is changing, so it is important to have this information well timed, not just as early as possible. Otherwise there could

be an information overload with the risk of misinterpretation and wasted resources.

This is a broad definition of TI. In its turn, TI can be understood as part of business intelligence. The present study will also regard TI as a task of integrated technology management.

TI systems

Several empirical studies have shown that in practice TI is pursued in a systemic way, a "casual" way, or a mixture of both. However, in order to be able to discuss different elements of TI, one can describe a **Technology Intelligence system** (without specifying the degree of systematization). In fact, there is a fundamental difference between "systemic" and "systematic." "Systemic" is a holistic understanding of elements which are related to each other (Daenzer 1976: 11). "Systematic," in comparison, describes a procedure. A management system can be visualized along the lines of Porter's (1985: 36) value chain. Since there is value creation throughout the **TI process** – i.e. need formulation, collection, analysis, dissemination, and application of relevant information – these activities can be interpreted as primary or direct activities of value creation. The value lies in improvements in decision-making; that is, when the quality (of content and timing) of information is improved in order to reduce uncertainty (Savioz 2001: 284). Supporting or indirect activities enable the primary activities. For a TI system, these enablers consist of:

- *TI management*: the basic management functions of a system are to design, direct, and develop the system.
- *TI mission and goals*: the TI mission and goals interact with the information needs, but there also is a direct link to the business mission and strategy.
- *TI structures*: the TI structure describes the arrangement of different elements of TI and the people involved.
- *TI tools*: TI tools include collection and analysis methods (for example, scenario analysis) and enabling infrastructure (for example, communication enablers like IT infrastructure).

These different activities are described below. The discussion is particularly enriched from the point of view of SMEs. However, it has to be noted that generally the focus in literature is not on SMEs, neither in conceptual research nor in empirical studies.

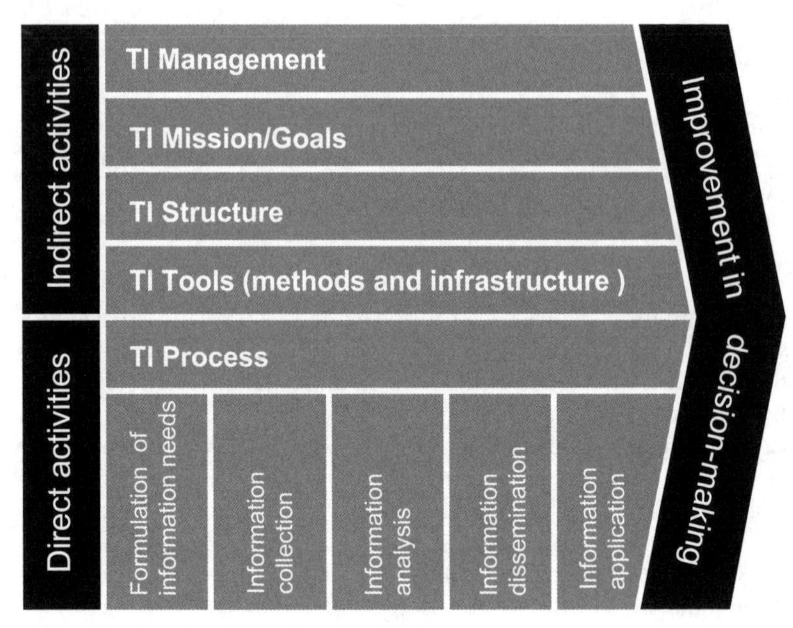

Figure 4.2 Direct and indirect TI activities

TI management

Much like an organization, a TI system should meet 5 conditions:

- single TI elements solve different problems independently, but blend well with other elements resulting in an integrated whole;
- single TI elements are reciprocally connected;
- a TI system is strictly an open system;
- there is a regularity in the TI system;
- a dynamic TI system can be guided and advanced.

Management of TI includes **designing**, **directing**, and **developing** elements of the system. Designing means generating a theoretical model, which reflects a "yet to be constructed reality." This is an "eminently creative process" (Ulrich & Probst 1988: 260). Directing is an "on-line" process constantly orienting the TI system so that it fulfills its mission and goals. Finally, developing the system involves consciously making shifts in order to cope with technological and social change. However, according to Rapoport (1988: 78), there is a constancy in change, which means a shifted system should be able to keep its identity and allow self-assertion. Thus, not all elements of the system need to be to be shifted. Within these "basic functions of management," implementation of the designed system happens radically (mostly after design or development) or incrementally (mostly during direction and development).

Since management is not a result or an element but a process guiding the system, it has a superior status. Nevertheless, it is an indirect activity of value creation because it enables direct activities.

There are few studies dealing with the management of TI systems from the point of view described above. However, there are some contributions in Ashton and Klavans (1997) dealing with "Managing Technical Intelligence Organizations in Business." All of them report on experience found in practice. The focus is on how to build up a TI system and on critical success factors. They can be summarized as follows:

- First, purpose is the essence of all effective intelligence operations. Thus, identification of those user needs at the beginning of a new science and technology (S&T) intelligence program or a new task for an established intelligence unit is critical.
- Secondly, a provision for adequate resources, particularly professionally trained and dedicated staff, should be assigned.
- Thirdly, no matter how it is organized, the intelligence system must possess the three basic intelligence operations: collecting and reporting, analyzing and forecasting, information processing and disseminating.

McDonald and Richardson (1997) describe in detail the design stage, the implementation stage, and the development and application stage of a TI system. These stages cope more or less with the three basic management functions. In the design stage, McDonald and Richardson propose interviewing management and other relevant personnel to obtain intelligence needs. This is followed by a prioritization of needs and of technological areas. Then resources should be assigned (i.e. monitoring and assessment specialists, a system manager function, as well as IT infrastructure). Lastly, the system should be organized. McDonald and Richardson differentiate between centralized, distributed, and hybrid options. In the implementation stage, the emphasis is on training and pilot operations efforts. Within the directing function, system modification and possible expansion are allowed. The development and application stages are described in several case studies. The conclusion is that the TI units or groups became more extended and involved during strategic planning.

Baisch (2000) examines the implementation of a TI system in more detail. He differentiates between implementation in a broad sense and that in a narrow view. The broad view of implementation is described as an ongoing and parallel process at different stages of management, which cope more or less with the above-mentioned management functions. The narrow view of implementation corresponds to the realization of the planned system. Even though the study concludes with an idealized implementation process of a TI system, it is difficult to derive practical recommendations. An important critique of this approach is that the TI system is considered in the abstract. This contrasts with Baisch's argument for parallel design and implementation of the TI system. No particular insight with regard to implementation of a TI system in an SME can be extracted from his study. However, it might be useful to retain some of his conclusions:

- The implementation of a TI system has to be supported autonomously by the superordinated system.
- The TI system has to be systematically integrated at every level of the company.
- Implementing a TI system should be based on formal and informal information exchange.

According to Krystek and Müller-Stewens (1993), there is no "patent formula" for implementation of an TI system. Implementation seems to depend very strongly on the organization's context, particularly on the "anticipative culture." Krystek and Müller-Stewens make a claim for a project-oriented implementation of TI with several project tasks. Some clusters of tasks are:

- *Development of a common TI language and understanding*: It is important that involved persons have the same expectations and understanding of a TI system. Preliminary discussions between initiators and the TI project team are necessary. Setting project goals and defining project measurements are part of these discussions.
- *Definition of the implementation project scope and duration*: The implementation project can be independent or part of a superordinated project (for example, a restructuring project, a strategy development project, etc.). The project can be implemented within a defined unit

or across the company. Furthermore, an overall time period should be defined for project phases. All this together can lead, for example, to a pilot project in a defined unit in the first stage, and a company-wide introduction of the TI system in a second stage.

- *Distribution of roles within the TI implementation project*: Tasks and responsibilities during the implementation project have to be defined. Special attention has to be given to the so-called "project champions." As in the case of innovations, implementation of new concepts, for example a TI system, needs champions. One can differentiate between champions with project-specific know-how (for example, knowledge in the field of TI) and champions with enough political power to push the project through.
- *TI system definition*: Any system details should be defined. In fact, Krystek and Müller-Stewens consider designing the TI system to be a part of the implementation project.

This discussion suggests that implementation and design of the TI system go hand in hand and can seldom be separated.

To summarize, the three basic management functions (designing, directing, and developing) seem to be valid, to some extent, for TI systems. Implementation is considered as one with the management functions. Only a few studies have examined the management aspect of a TI system. There are gaps in the literature in this area; SME concerns are particularly neglected.

TI mission and goals

The TI mission and goals define the purpose, and therefore the required output, of the TI system. This includes several aspects which will be discussed below.

The TI mission always should be related to **the business mission and strategy**. This link is confirmed by an empirical study conducted by Jennings and Lumpkin (1992: 800), but the study does not ascertain whether caution runs from strategy to intelligence activities or vice versa, or in both directions at once. This phenomenon is a veritable chicken–egg problem: "Who drives whom?" Abell (1999: 80) describes that different strategies, for example the dual strategy consisting of a "today-for-today" and a "today-for-tomorrow" orientation, demand different intelligence activities. Thus, when defining a TI mission and goals, this idea should be taken into account. One approach to the chicken–egg problem could be to say that today's strategy tends to drive short-term oriented TI, while long-term oriented TI is necessary to drive tomorrow's strategy. In an adaption of Mintzberg (1995: 15) and Simons (1995: 154), Jung and Tschirky (2002: 15) suggest installing a so-called "interactive technology control system," in which intelligence plays a major role.

Furthermore, Hambrick (1982: 169), on the other hand, could not find a direct link between strategy and TI, but observed a general lack of awareness of organizational strategy among employees and a difference between how the researchers and the executives conceive strategy, both of which could be causes of failing intelligence activities. Thus, TI activities always should be based on a TI mission and goals, which are linked to the business mission and strategy.

The literature posits different **aims of conducting Technology Intelligence**. (A few examples are given in Table 4.1.) In contrast to such objective-oriented views, Peiffer (1992: 106) proposes a differentiation between an "**inside-out**" and "**outside-in**" **perspectives**. The first focuses on observation of technologies within the existing area of action. The latter is an unbiased observation of general technological trends (see Figure 4.3).

Table 4.1 Different TI aims

Ashton & Klavans (1997: 11)	*Reger et al. (1998: 5)*	*Lang (1998: 75)*
To provide early warning of external technical developments or company moves that represent potential business threats or opportunities	To expand their present business through technological improvements	The spectrum of functions and technologies is known; observe development and changes within these fields
To evaluate new product, process, or collaboration prospects created by external science and technology activities in time to permit appropriate responses	To generate new technological knowledge for the development of new fields of business	Observation of technologies on the basis of particular spectrums of functions
To anticipate and understand science and technology related shifts or trends in the competitive environment as preparation for organizational planning and strategy development	To anticipate technological discontinuities and global changes, so that the firm is not taken unaware or "submerged" by new paradigms or new competitors	General observation of the technological environment for any opportunities

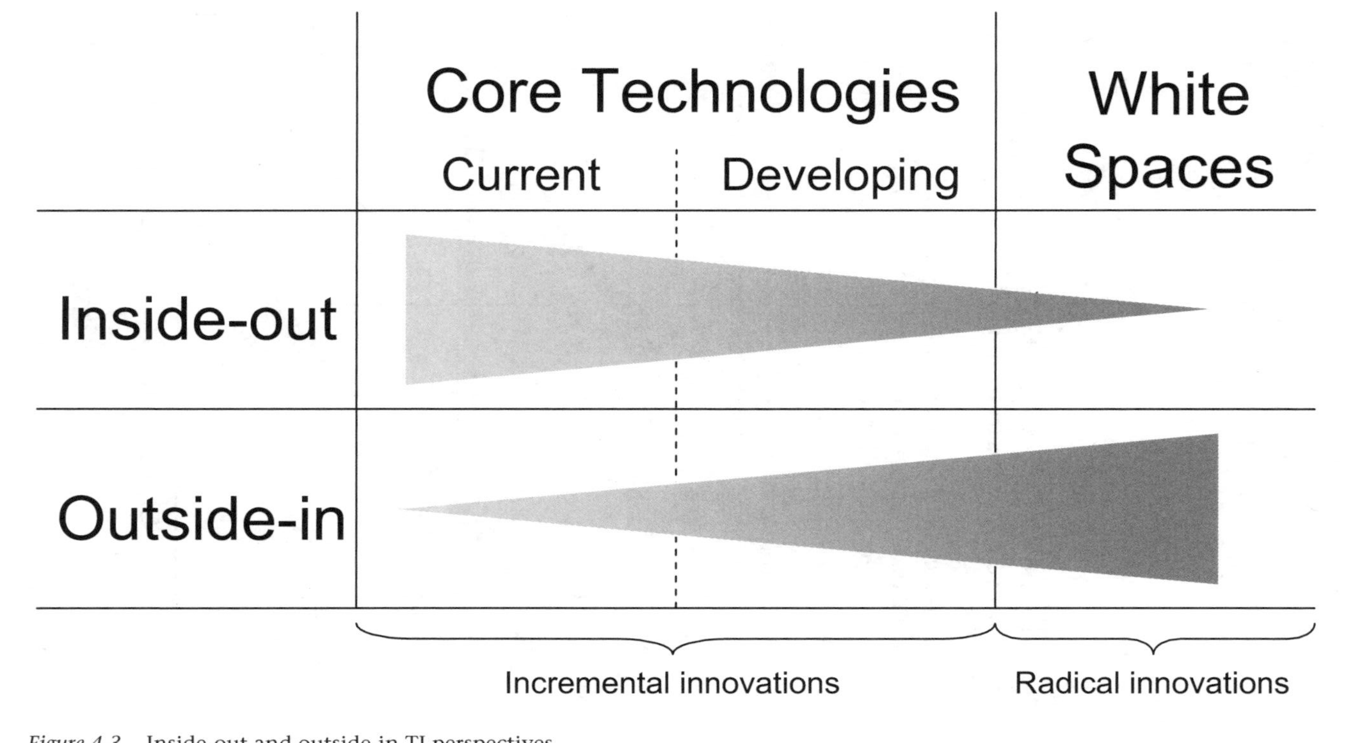

Figure 4.3 Inside-out and outside-in TI perspectives
Source: adapted from Lichtenthaler 2000: 33.

A combination of these reflections can be represented as an observation matrix (Figure 4.4). The observation area always has a time and a content dimension. Depending on the strategy, the organization puts the focus on a particular observation area. While "followers" seem to place their focus of interest on technologies and developments in the present or the near future, and in existing or familiar businesses ("keeping abreast area"), "leaders" additionally seem to pursue observation beyond this limitation ("looking beyond").

Aguilar (1967: 18) does not describe different observation areas, but differentiates between four **observation modes**:

- *Undirected viewing* is defined as general exposure to information where the viewer has no specific purpose in mind with the possible exception of exploration. This mode is characterized by the viewer's

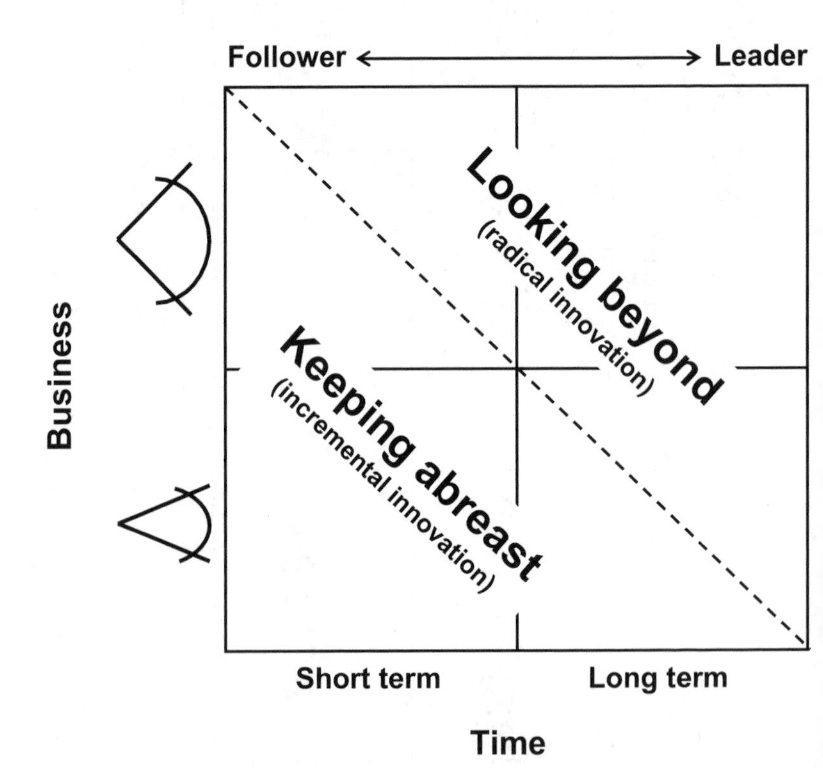

Figure 4.4 "Keeping abreast" versus "looking beyond"

general lack of awareness as to what issues might be raised. The sources of information are many and varied, the amounts are relatively great, and the screening is generally unrefined.

- *Conditioned viewing* is defined as directed exposure, not involving an active search, to a more or less clearly identified area or type of information. It frequently serves to signal a warning or to provide a cue that more intensive scanning should be instituted. Conditioned viewing differs from the undirected type principally in that the viewer is sensitive to particular kinds of data.
- *Informal searches* are defined as relatively limited and unstructured efforts to obtain specific information. They differ from conditioned viewing principally in that the information wanted is actively sought.
- *Formal searches* refer to deliberate efforts – usually following a preestablished plan, procedure, or methodology – to secure specific information.

Common to all these facets is the fact that the **primary mission of TI systems** or programs is to provide information that supports business decisions. Furthermore, TI results do not simply trigger decisions, but increase the scope of these decisions.

No literature on SMEs deals explicitly with the aspects I have discussed above. Empirical studies on this subject mostly focus on Fortune 500 companies or the equivalent. Nevertheless, insight from some studies conducted on SMEs, examining planning activities and environmental uncertainty, may give a better understanding of the strategy/intelligence relationship. However, the results can be contradictory: Yasai-Ardekani and Nystrom (1996: 201) found that small as well as medium-sized and large organizations were able to develop effective scanning systems, whereas Smeltzer et al. (1988: 57) argue that small firms cannot afford the luxury of specialized scanning systems. This view is supported by Beal (2000: 31), who argues that SMEs often miss opportunities because they lack the financial resources to achieve them. Mohan-Neill (1995: 19) found that when smaller firms make decisions they utilize less marketplace information than do larger firms. These authors argue that SMEs engage in less formalized planning than larger firms. Yet another study rejects this view by concluding that there is no difference in scanning with regard to the company's size (Lyles et al. 1993: 44). Matthews and Scott (1995: 48) observes that sophistication of strategic and operational planning in SMEs declines with increasing environmental uncertainty. Finally, Dou

(1995: 104) specifies that TI cannot be done efficiently within SMEs if there are no strategic reflections on the firm's business, its future, and product evolution.

These controversial findings do not allow for presumption about whether an SME should or could practice intelligence activities, and if there is an obvious link between strategy and intelligence activities. It is not clear if conclusions drawn from large firms can be transferred to SMEs. In addition, there are no prescriptions with regard to the explicit goals or the observation focus of intelligence activities in SMEs. Thus, there is an urgent need for new concepts.

TI structures

TI structures describe how TI activities are assigned to different units and people, and how they are organized. This chapter has two parts: the first part discusses various aspects of hierarchical characteristics of a TI system and models of how to realize TI activities. The second part deals with the roles and people involved. Both parts begin with general insights from the literature, and end with a short discussion about SME concerns.

Hierarchical characteristics of a TI system

The literature on TI generally examines (in empirical studies) and proposes (in conceptual works) solutions with a hierarchical view. A strong emphasis has been put on the distribution and coordination of different tasks between the corporate level and divisions or business units. Most solutions are based on a formalized model, consisting of a centralized TI unit at the corporate level and some additional decentralized TI elements. An overview is given in Figure 4.5 (Lichtenthaler & Tschirky, 2001). Some elements are within the company, others are external.

Formal structures hide the danger of intelligence unilaterally benefitting intelligence for top management. Lichtenthaler (2000: 291) classes this type as "**intelligence of the organization**," as opposed to "**organizational intelligence**," which takes into account that in reality decisions are made throughout the company at any level within a certain strategic framework (Figure 4.6, adapted from Lichtenthaler 2000: 292). Thus, this type of TI goes beyond just supporting decision-making, in that it enhances organizational learning: "the [TI] function may often only serve as a facilitator of learning with an unclear direct implication on decision-making or planning" (Hedin 1993: 132).

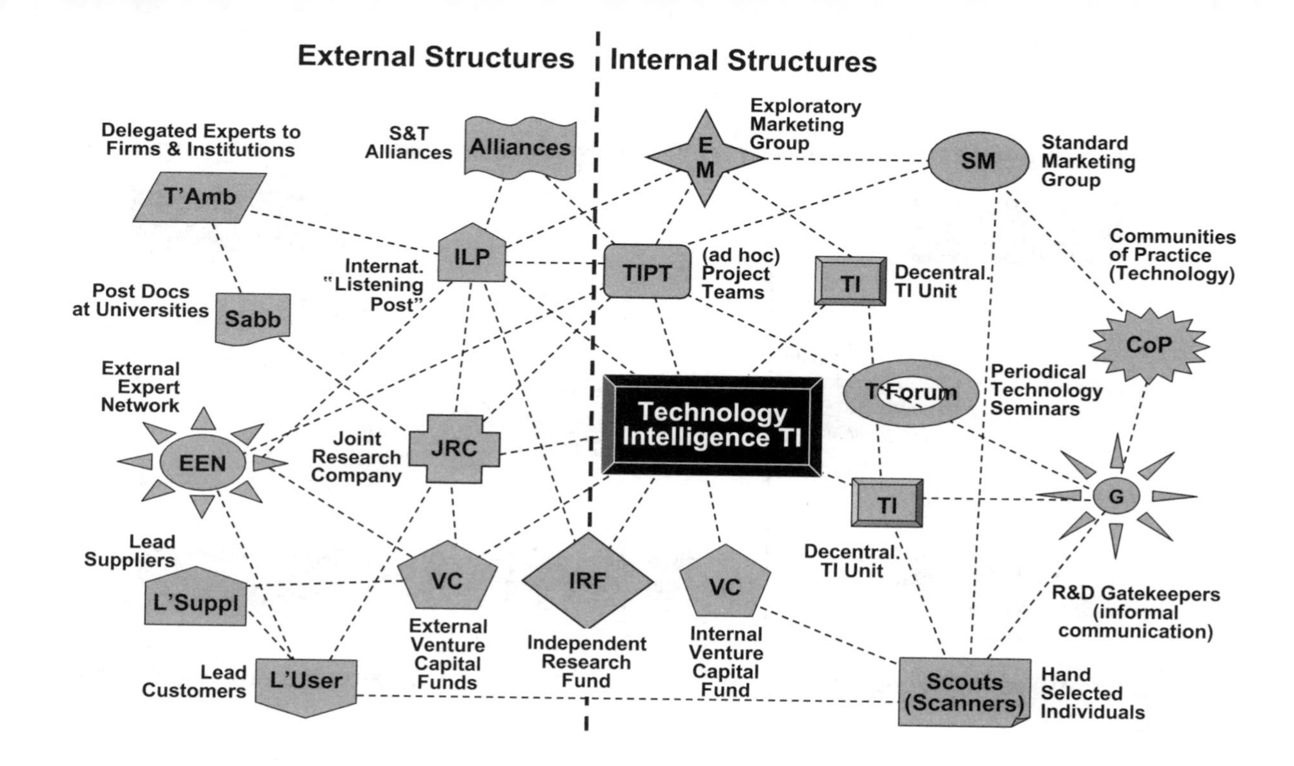

Figure 4.5 Internal and external technology-intelligence structures (Lichtenthaler & Tschirky 2001)

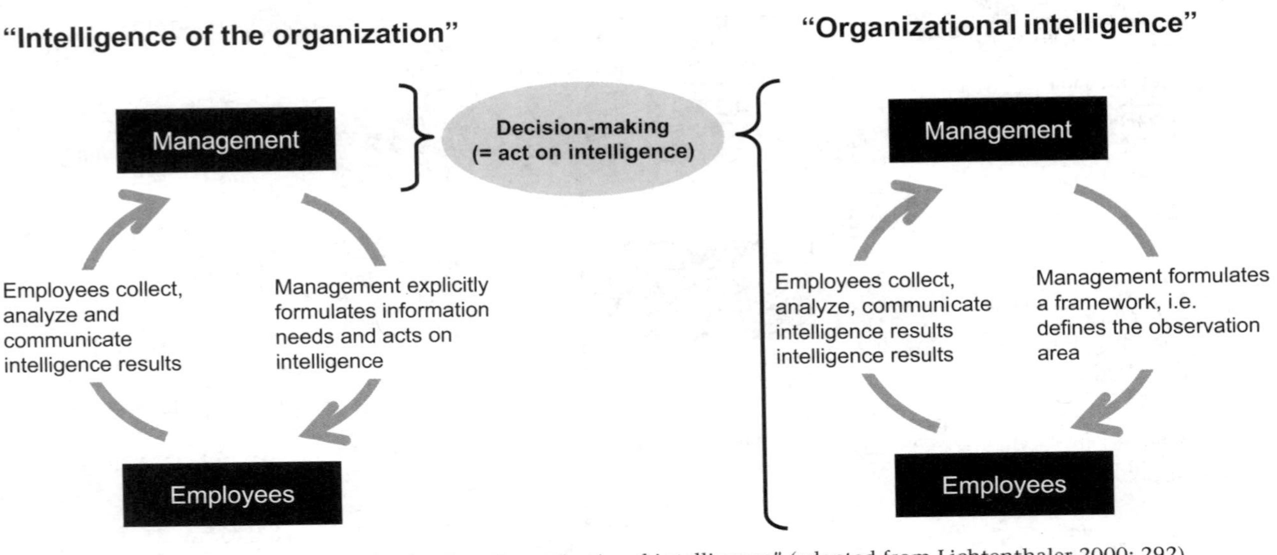

Figure 4.6 "Intelligence of the organization" vs. "organizational intelligence" (adapted from Lichtenthaler 2000: 292)

In addition to a formal and hierarchically structured organization of TI, TI activities can be executed informally or as projects. Formal and informal structures are run continuously and projects are discontinuous. Lichtenthaler (2000: 248) differentiates between three types of coordination of TI activities: structural, hybrid, and informal coordination (see Figure 4.7, adapted from Lichtenthaler 2000: 248). He observed in his study that all of these types of coordination are run in parallel in most companies. A problem could be the fact that formal and informal activities are sometimes at odds.

These diverse possibilities of distribution and coordination of TI activities show in some way that anything is possible (and indeed practiced). The existing literature does not provide a consistent idea of how to organize a TI system. Most authors describe solutions based on hierarchies or organizational charts. Nevertheless, several authors describe factors that influence the organization of a TI system:

- *Company culture*: if the intelligence program does not fit with the management culture of the company, it might be viewed as threat to existing interests. Results from the program might be objected to or ignored. In addition, the success of intelligence activities probably depends on good communication, which in turn depends on a healthy company culture.
- *Technology life-cycle*: while uncertainty in early stages of technologies demands a broad TI platform with a learning aspect, it is recommended to organize a structured TI for mature technologies, i.e. when the potential for competitive differentiation is high. For very mature technologies, the focus of TI is on market aspects.
- *Basic company structure*: depending on existing structures, and on the degree of diversification, TI activities can be assigned to different units.
- *Innovation strategy*: depending on the degree of innovation (incremental or radical innovation) and on the positioning (technology leader or follower), different persons and units are involved, and thus the organization of TI might be different.
- *Decision-making process*: while the existing TI literature prefers to describe "intelligence of the organization" approaches in the case of a top-down decision-making process, Lichtenthaler (2000: 349) observes in his study that the competence of decisions in most firms is decentralized at lower levels (bottom-up). This affects TI organization.

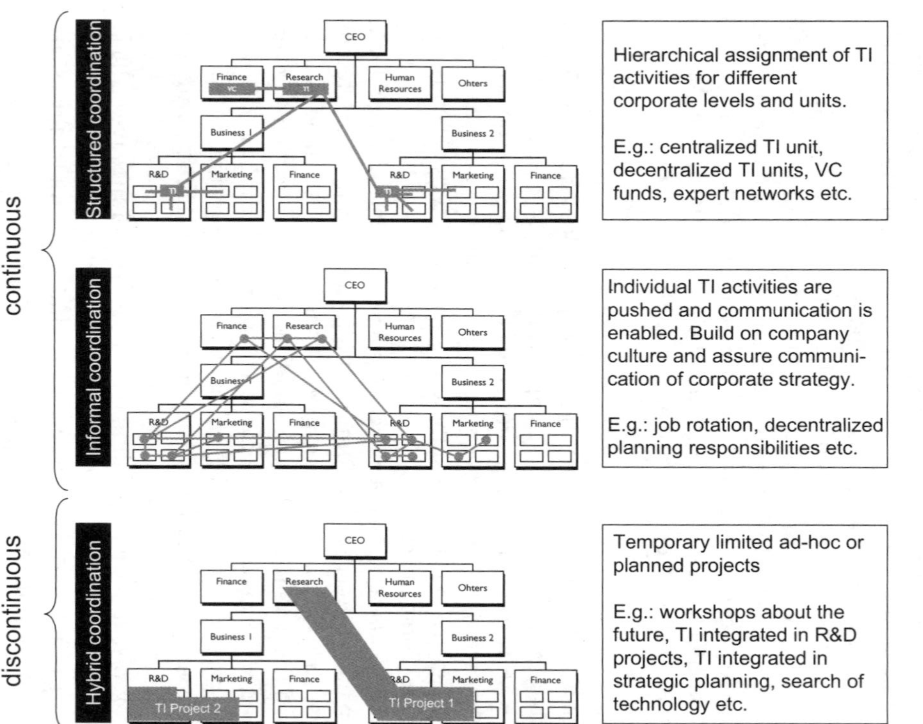

Figure 4.7 Structured, informal, and hybrid coordination of TI activities (adapted from Lichtenthaler 2000: 248)

- *Industrial sector*: information in science-based sectors is mostly available from general science, but the implications are difficult to interpret; whereas in technology-based sectors, technology disruptions often occur from outside the sector. This differentiation implies a different organization of TI.

In the literature, the discussion of TI structures and influencing factors is based on studies which are focused on multinational, diversified companies, or in which no specification is made. However, since most authors discuss TI at different levels of the organization, for example at corporate levels or divisions, these solutions are implicitly significant for large companies. Some authors point out a certain dependency on the TI system and company size, but proposals are missing for the TI structures and influencing factors that are valid for SMEs.

An interesting type of TI may be an intercompany or industry-wide organization. However, intercompany TI systems did not find an application in practice. Hassid et al. (1997: 129) compare some industry-wide advances in France (for example, Agence pour la diffusion de l'information, ADIT [the "Information Agency"]), the United States (for example, the Manufacturing Extension Partnership, MEP), Japan (for example, Kosetsushi), and Germany (for example, Steinbeis Foundation). All of them more or less take the role of information provider. Thus, Krystek and Müller-Stewens (1993: 152) state that such industry-wide systems should just be understood as complementary to the company's own TI system.

People and roles in a TI system

A discussion about the problem-owners of different activities might help to understand potential parallels to SMEs. Several authors name different roles within a TI system; these are summarized in Table 4.2. Such roles demand various competencies. Normally each role is occupied by one or more persons in large companies. However, it is difficult to find people having all of the necessary competencies. Jakobiak and Dou (1992: 9) suggest that the chief executive and some engineers should take care of TI activities within SMEs.

Neither empirical studies nor conceptual works examine in detail the different roles within an SME's TI system. It is not clear whether roles should be the same as for large companies or not. There seems to be a need for a more detailed understanding of what kind of roles are assigned to whom within SMEs.

Table 4.2 Roles within a TI system

Krystek & Müller-Stewens (1993: 259)	*Lichtenthaler (2000: 258)*	*Kobe (2001: 196)*
Facilitator: she/he supports the TI system with regard to the top management, most notably during the designing and implementation stage. She/he is typically her/himself a member of top management or member of the board.	**Process coordinator**: she/he is responsible for designing and improving the TI system. She/he coordinates TI activities and promotes the TI spirit within the company. In American companies, she/he was typically a CIA or NSA specialist.	**Idea medium**: her/his task is to trace various ideas within the company in order to sort and analyze them. She/he can do this voluntarily or can be mandated.
Expert: she/he has specialized knowledge to manage the TI system. She/he achieves a coordination function within the TI system and participates in operational TI activities.	**Information specialist**: she/he seeks information on demand. Typically she/he has specialized skills in database research.	**Process promoter**: she/he is responsible for linking technological information and product innovation. This should be a permanent task.
Scanner: she/he is a TI worker who first of all collects information. She/he might be involved in analysis activities. She/he has specialized knowledge in database research.	**Analyst**: she/he analyzes the gathered information. She/he typically has outstanding technological and communication competencies.	**Expert**: she/he is assigned to observe technologies. She/he collects and analyzes relevant information. She/he might be supported by specialists in information collection.
External: for any activity a company might revert to an external expert. This is typically true for collecting and analysis activities.	**Method specialist**: on the one hand, she/he applies different methods like scenario analysis; on the other hand she/he helps others to use them.	

The TI process

The TI process consists of different stages in the value-creation process of intelligence. Three major conceptions of this process are as follows:

- **Peiffer's concept** (1992): Peiffer differentiates between two major steps: identification & observation and assessment of prospected

technology fields. He uses the "twirl–aspirate–filter" metaphor. This means identification of all relevant technology fields in a very broad environment, followed by several assessments. Rejected technology fields are "recycled" in further assessments. The concept's focus is on the identification of *the* right technology. However, it does not allow for *any* technological information, i.e. changes in employed technologies. In general, this concept reflects an "outside-in" view.
- **Ashton et al.'s concept** (1991, 1994): Instead of a broad and explorative view, they adopt a customer-related view of TI. Intelligence activities are planned, so that a certain intelligence need is expressed. A step-by-step process follows, including collection, analysis, dissemination and application. The feed-back step shows, that this process must be formalized. In general, this concept reflects an "inside-out" and "top-down" view of TI of the organization.
- **Lichtenthaler's concept** (2000): Lichtenthaler observed in his study that these idealized concepts do not reflect the companies' reality. He observed several parallels between different stages of intelligence activities, for example analysis activities already occurring in the collection stage, and communication being important at any stage. Thus, the limitations of each stage are very blurred, and the stages live in parallel. Therefore, the process needs to be broken down into an information extension stage (conveyance) and an information reduction stage (convergence) of TI. The underlying understanding is the idea of organizational intelligence, which implies participation of all employees (or at least more than top management) in all activities of the TI system, especially in application. In fact, employees do TI activities within a certain common framework, which changes from one to another transition of the conveyance and convergence stages. Depending on the focus of intelligence activities, the time horizon of this transition may change. In general a traditional, sequential view of the process is not unnecessary, but it solely reflects the production (and therefore value creation) process of intelligence.

These three concepts show some differences of understanding of the TI process. However, they are complementary. Literature does not provide insight into the practical use of TI processes in SMEs. For this book, the following aspects are retained:

- The TI process is a value creating process. It is not a step-by-step process, but a parallel assembly of diverse interacting TI activities.

- The activities of the TI process are: formulation of information need, information collection, information analysis, information dissemination and information application.
- The barriers between different TI activities are blurred.
- There is an explicit or implicit formulation of information need. Information need can emerge at any level of the company.
- The "outside-in" logic, i.e. scanning, and the "inside-out" logic, i.e. monitoring, are complementary.
- The produced intelligence is applied at any level of the company.
- The interaction between TI activities depend on the nature of need, time horizon, competencies and company culture.
- Finally, the TI process corresponds to the paradigm of "organizational intelligence."

Always keeping in mind the iterative and parallel interaction of TI activities, a short description of each follows. Methods are an important part of some stages. However, they are described in a separate chapter because some of them cannot be assigned to a specific stage, for example patent analysis is used in the collection as well as in the analysis stage.

Formulation of information need

The scope of the formulation of information need is of two types: firstly, information need should give the impulse for other TI activities, secondly, it prevents an information overload by limiting the observation area, respectively limiting the use of resources. Both the impulse and the observation area are in close relation, even partially overlapping with TI mission and goals.

On the one hand, the impulse may be manifested by an **explicit formulation** of needs for information, mostly from a top-down initiative, in the form of a specific demand or a general declaration of requirements. In this case, one can speak about an input in the TI system, and consequently TI activities will be reactive to the explicit needs. On the other hand, needs may appear **implicitly**, i.e. by means of emergent strategies. In this case, TI activities will be based on an implicit impulse. This is particularly true from the point of view of "organizational intelligence."

An objective and precise formulation of strategic information need does not seem to be possible. Thus, a mix between an "outside-in" and an "inside-out" perspective helps to successively reduce the amount of information within a certain framework, which is based on TI mission. The emphasis of each perspective depends on time

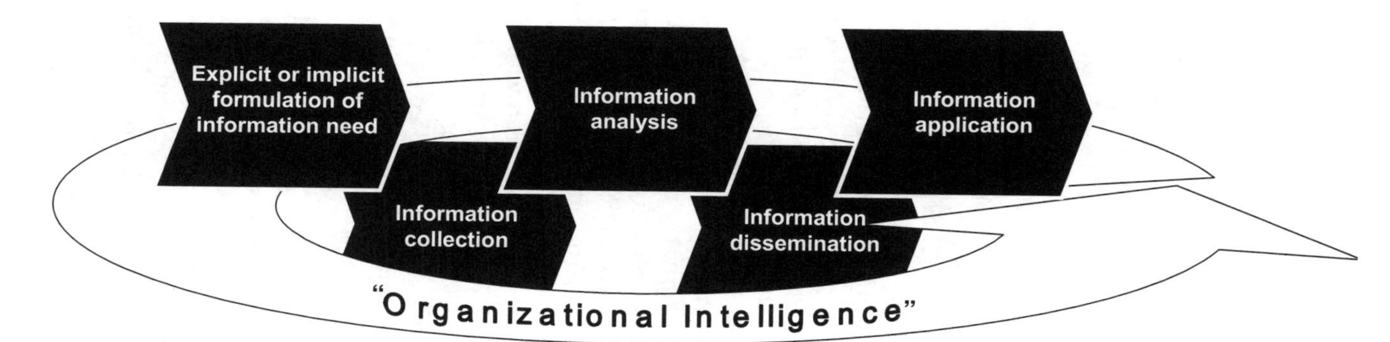

Figure 4.8 Representation of TI activities in the TI process

horizon, industry, and technology planning and strategy. General functional thinking is promising, in contrast to thinking in tight technology categories, in order to avoid blind-spots. These two scopes are illustrated in Figure 4.9.

Anyhow, the existing literature does not provide a clear framework for defining information need. This might be because of its industry and strategy dependency. In addition, no study has been conducted to examine the formulation of information need for SMEs.

Information collection

Determining the scope of information collection is the first stage meeting in the information needs described above. Two important questions come up: Who does it, and what are the information sources?

The assignment of the collecting task depends on competencies. "Competence is an ability to sustain coordinated deployment of assets in a way that helps a firm achieve its goals. Here we use the word ability in the ordinary language meaning of a power to do something" Sanchez et al. (1996: 8). One can distinguish between internal and external technology competencies. The more the observation field matches with internal technology competencies, the more intensely changes are perceived. In turn, the capability of getting information decreases the more the observation focus clashes with internal technology competencies. This can be justified by describing the necessary previous knowledge or experience of individuals within the company and their ability to anticipate the appropriate information sources. These people are typically researchers and engineers with a lot of experience. In the case of missing technology competencies, (internal) TI specialists and external experts may fill in the gaps.

In literature, a big emphasis has been placed on information sources. A distinction between **formal and informal sources** seems to be appropriate. Some information sources are represented in Figure 4.10 (adapted from Keller 1997: 50).

The existing literature does not provide substantial insight into who uses which sources in which cases. Principally all formal and informal sources are accessible for SMEs. Fann and Smeltzer (1989: 43) found that SME managers collect information essentially from suppliers/vendors, customers, employees, and periodicals. They use patent databases mainly to acquire information for legal purposes, not for technical data because of the cost in terms of personnel time and expertise. In addition to lacking human resources, available

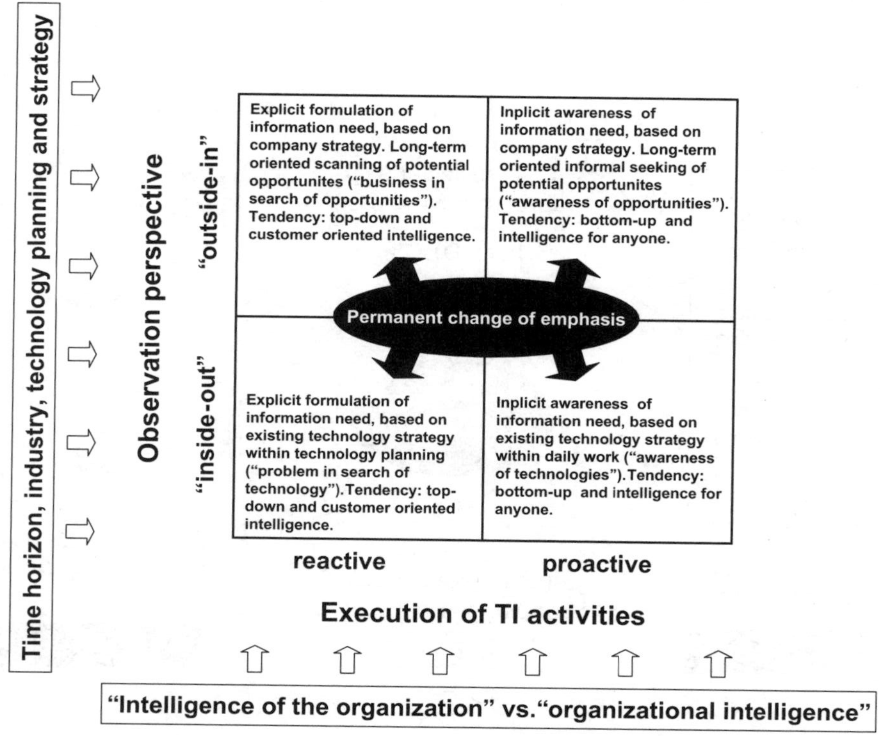

Time horizon, industry, technology planning and strategy

Observation perspective

	reactive	proactive
"outside-in"	Explicit formulation of information need, based on company strategy. Long-term oriented scanning of potential opportunites ("business in search of opportunities"). Tendency: top-down and customer oriented intelligence.	Inplicit awareness of information need, based on company strategy. Long-term oriented informal seeking of potential opportunites ("awareness of opportunities"). Tendency: bottom-up and intelligence for anyone.
"inside-out"	Explicit formulation of information need, based on existing technology strategy within technology planning ("problem in search of technology"). Tendency: top-down and customer oriented intelligence.	Inplicit awareness of information need, based on existing technology strategy within daily work ("awareness of technologies"). Tendency: bottom-up and intelligence for anyone.

Permanent change of emphasis

Execution of TI activities

"Intelligence of the organization" vs. "organizational intelligence"

Figure 4.9 Observational perspective vs. impulse for TI activities

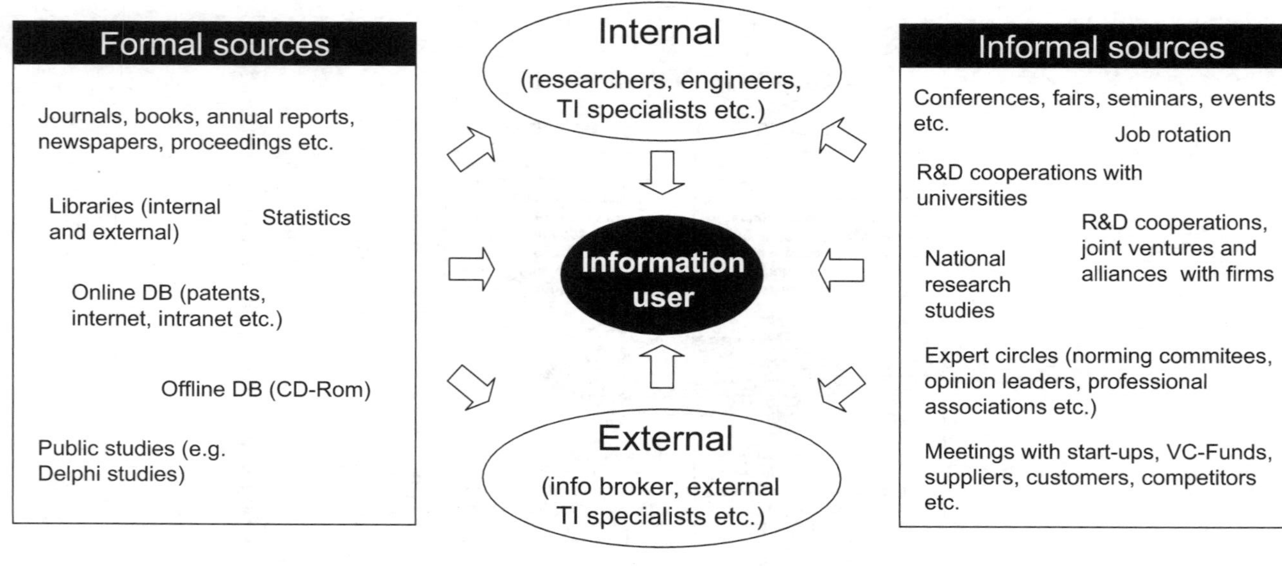

Figure 4.10 Formal and informal information sources

financial resources limit access to formal sources. Thus, Lybaert (1998: 188) suggests cooperating with external consultants. Furthermore, forming links with other organizations is a mechanism by which SME can reduce the problem of extending their knowledge-base. Nonetheless, Curran et al. (1993: 24) found that SME manager's contacts are much more limited and much less than notions such as "network" and "networking" imply.

To summarize, the general results on the "who does it" question seem to be valid for companies of any size. The limitation may be that internal technology competencies are more restricted for SMEs than in large firms. Concerning the use of information sources within SMEs there are some controversial findings in literature. This does not allow for making a solid statement about the use and usefulness of different information sources for SMEs.

Information analysis

The goal of analysis is to give the information meaning. The degree of analysis effort depends strongly on the clarity of information and on the desired insight. Some collected information already comes to the decision-maker in an appropriate and interpreted form. Thus, no analysis is necessary. In contrast, there may be only very fragmented pieces of information which require a lot of effort for analysis. Therefore, the use of different analysis tools and the decision by whom analysis should be done depend on the context, for example on technology strategy (for example leader vs. follower) and environment complexity.

Analysis has three different functions: **filter, integration and assessment**. The filter function is to reduce the quantity of information by checking the relevance of the information to the company, and to assess the quality of information. This function is partly fulfilled implicitly in the collecting stage. The integration function is expected to integrate information in the company context, which requires appropriate background knowledge. Then, the assessment function is to estimate the strategic meaning of information to the company. It is in this stage that information becomes intelligence.

Very important for analysis are analysis tools. Thus, they are discussed in a separate chapter describing TI tools. However, tools and methods cannot replace human judgment. Assignment of analysis activities to individuals or groups is therefore crucial. Some, rather technical tasks, for example a trend extrapolation, need individuals' special skills. Other tasks do not need skills but intuition, for example

Delphi studies. Since intuition is not about the right or wrong answer, but about plausible or implausible arguments, intersubjectivity of multiple individuals seems to be the next best solution.

All of these findings can be transferred to SME, with the same restriction as for information collection: restricted internal competencies, in particular, restricted know-how in use of analysis tools, limit the range of action. In conclusion, it is recommended that SMEs revert to external experts for some analysis activities.

Information dissemination

Information dissemination is the stage where intelligence reaches potential users. An important topic is the communication (or information-flow) logic. A strong emphasis is placed on **information push vs. information pull**. The existing literature does not provide a complete answer for an optimal logic. Ebadi and Utterback (1984: 582) and Allen (1986: 112) found in empirical studies that increasing frequency of communication increases innovation success, which is favorable for SMEs. On the other hand, they found communication channels that too formalized repress innovation success.

Certainly the information-flow logic depends on the intelligence organization: if there is an "intelligence of the organization," i.e. the user (mostly top management) is known, information should be pushed, whereas "organizational intelligence" needs information push and pull at the same time. Krystek and Müller-Stewens (1993: 14) make a claim for communication by participation. Information's value increases the more people use it. But there is a danger of outflow of strategic information to competitors because of an uncontrolled information flow. Thus, several companies protect their information rigorously with "counter-intelligence" measures.

Some authors argue with Daft and Lengel's (1986: 554) "Media Richness Theory," which says that the more complex the communication situation is (for example, time-critical information), the more complex the communication medium has to be (for example, face-to-face communication). Lichtenthaler (2000: 342) observed in his study a very broad spectrum of communication media and that no communication media is systematically preferred for any situation. However, he observed that indirect communication by means of formal "Technology Intelligence products," like long reports, do not appear often. Companies prefer tailor-made communication, adapted to the user and depending on the decision to be made. Of course IT plays an important role in communication, in any case. This aspect is also discussed in the chapter about TI tools.

There is surprisingly little literature on communication in SMEs. Since communication is an important aspect of innovation, one could expect to find insight from innovation and knowledge management in SMEs. However, there are just simple statements like "there is direct communication" and "there are short communication distances." These facts certainly stimulate information exchange, but questions on communication between "employees and owner," "researchers and engineers," "individuals and the group," etc. are not answered.

Information application

"The use stage is where the intelligence efforts pay off" (Ashton & Klavans 1997: 14). The value of the generated intelligence can be realized through direct facilitation of decision-making and through learning.

Direct facilitation is through delivering appropriate information to the customer (mostly top management), which copes with explicitly formulated information needs. Gerybadze (1994: 133) rates this unilateral view on its own rather than as a traditional intelligence paradigm. He makes a claim for interactive decision-making, so customers could be anybody. Krystek and Müller-Stewens (1993: 195) see intelligence as having an added value particularly if it has been generated collectively by decision-makers, because there is an organizational learning component which increases competence and the pace for further decisions. This extended view of the application of TI helps a company to improve planning assumptions, to improve portfolio management, to enhance decision-making ability, to improve the process of selecting research projects and allocating resources to them, and to increase awareness of threats from unscrupulous competitors.

Nevertheless, Bryant et al. (1997: 183) state that TI is just one of the many tools that planning/decision-making utilize. In addition to intelligence, decision-makers often act intuitively, sometimes even irrationally ("Rational actor" vs. "Political actor"). It happens that they just (mis)use TI to legitimize decisions already taken.

There is no literature specifying decision-making support in SMEs. However, the insights above seem to be applicable to them. Since there is little literature on how information is generally applied, further research on this is necessary.

TI tools (methods and infrastructure)

Until now, the discussion of TI has been limited to content and people. Just a few words have been said on tools that support TI activities. In this book, tools are understood as **TI methods** and **supporting technical infrastructure**.

TI methods

Existing studies do provide a wide range of the methods which can be used in TI. A central question is: which method should be used in which case? Because methods are used in various stages of the TI process, mainly collection and analysis, no specification is made here.

It has already been mentioned that influential factors of application may be technology strategy and environmental complexity and uncertainty of the industry. Another influential factor is time focus. Most authors differentiate between **quantitative and qualitative methods**. Figure 4.11 gives an overview of some TI methods in relation to time focus and characteristics.

Another criterion of choice seems to be the complexity of the method itself. Instead of the most exact method, it is recommended to use the most comprehensive method. Thus, the extension of the sophistication of methods is only possible if most involved people support the extension, not because "method fetishists" push it. In addition, the choice of several methods implicate application of other methods (for example, various methods are used to generate scenarios) and should therefore be understood as compounded methods. Lichtenthaler (2000: 330) observed a shift from quantitative to qualitative methods within the last few decades. He explains this with the new paradigm of TI, in which activities are participative rather than delegated to special units. In addition, he examined the application of TI methods in three industries. Table 4.3 shows some TI methods in relation to supported functions and their use in these three industries.

The use of TI methods in SMEs is seldom examined, i.e. whether methods generally designed for large companies can be applied in SMEs or not. In general, most methods seem to be too sophisticated and too expensive for application within SMEs. Thus, some authors try to break down "big" tools to SME-specific solutions, for example roadmapping. Another solution might be to revert to national studies for information collection and to external experts for analysis. Regardless, the learning component of TI would get lost.

Technical infrastructure

"The key to a successful implementation of competitive information systems is facilitating the systematic collection and distribution of intelligence information. This process is often defined as computerizing the

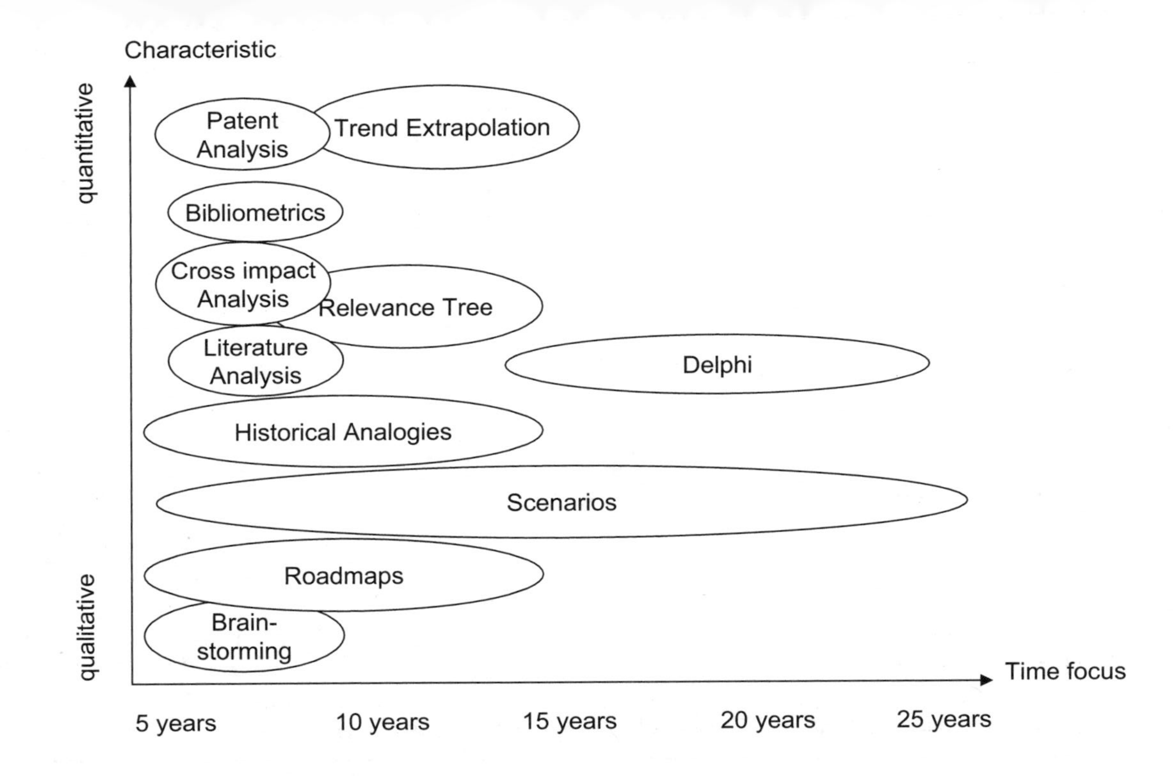

Figure 4.11 TI methods (Lichtenthaler 2000: 41)

Table 4.3 The use of TI methods in different industries and supported functions

TI method	*Information reduction*	*Organizational learning*	*Exploration*	*Communication*	*Comparison and control*	*Pharma*	*Electronics*	*Machinery*
Publication frequency analysis	×		×		×	often	sometimes	rarely
Bibliometrics	×		×		×	often		
Quantitative seminar observations	×				×	sometimes	often	rarely
Patent frequency analysis	×		×		×	sometimes	often	often
Patent linkage analysis	×		×		×			rarely
S-curve analysis	×							
Benchmarking studies	×				×	often	often	often
Portfolios	×	×		×		often	often	often
Delphi studies	×		×		×			
Expert panels	×				×	often	rarely	sometimes
Flexible expert interviews	×				×	often	often	often
Technology roadmaps	×	×	×	×		sometimes	often	
Product-Technology roadmaps	×	×	×	×			often	rarely

Table 4.3 The use of TI methods in different industries and supported functions – *Continued*

TI method	*Information reduction*	*Organizational learning*	*Exploration*	*Communication*	*Comparison and control*	*Pharma*	*Electronics*	*Machinery*
Product roadmaps	×	×	×	×		often		
Learning curves	×						often	sometimes
Simulations	×		×			sometimes		
Option-pricing methods	×					sometimes		
Scenario analysis	×	×	×	×		often	often	often
Lead-user analysis	×	×	×	×			often	sometimes
Quality function deployment	×	×	×	×			sometimes	often

process" (Hohhof 1997: 262). Over the past few years the potential of **information technologies** (IT) has grown enormously, and with this, the potential support in the Technology Intelligence process. IT supports the process at most stages (collection, analysis, dissemination), except for need formulation and intelligence application. Thus, IT remains a supporting tool which facilitates several tasks, but does not replace human thinking. Computer support roles can be listed as follows:

- provide access to secondary information for both intelligence systems analysts and intelligence system users,
- identify and distribute primary information,
- organize information for retrospective retrieval and provide access to other internal information sources,
- facilitate the intelligence analysis process, and
- distribute intelligence products to system users.

The scientific literature does not discuss IT use in the overall TI process. There are some contributions on how IT supports collection, analysis, storage, and dissemination isolated from the overall process. However, Guimaraes and Armstrong (1998: 50) found that companies with an above-average information system (IS) support perform above-average intelligence effectiveness.

In contrast, there are numerous contributions on IT solutions for intelligence in the popular press. However, most of them equate IT solutions, like Data Warehouse and Data Mining, with "business intelligence." They are mostly a kind of advertising for IT products. A detailed discussion and consideration of various IT products and solutions would exceed the scope of this book.

The literature suggests that computer usage by small businesses is largely restricted to administrative tasks. However, Chen and Williams' (1993: 98) study of small English firms did note that there is also some evidence that the main use of computers is moving from record-keeping (word processing and bookkeeping) to decision-making (financial modeling and data management). Medium-sized companies tend to use the various software packages more extensively than small firms (Bridge & Peel 1999: 83). A recent study examining the use of IT (especially use of networks etc.) in SMEs could not be found.

Thus, IT infrastructure is accepted in this book as a supporting tool for various stages in the TI process. The kind of hardware and software which should be used to support each stage of the intelligence process in SMEs should be examined further.

Improvement in decision-making (TI measurement)

Since the application of intelligence is both direct facilitation of decision-making as well as learning, the value added from the TI process and its enablers is not trivial. In fact, learning means indirect facilitation of further decision-making at the same time. The literature is not clear on this point. While Eger (1995) states that successful intelligence programs must also possess the ability to measure the "monetary value" of its operations, Kahaner (1997: 231) says that effects of intelligence activities are indirect and cannot be measured.

However, one can determine the difference between **effectiveness and efficiency** of TI activities. The effectiveness can be surveyed with a subjective measurement of customer (decision-maker) satisfaction and with objective "business cases" on time and money used, and on sales. The first is possible if TI follows an "intelligence of the organization" paradigm on which TI activities are started by a reactive impulse and by one single user (mostly top management), but not with the "organizational intelligence" paradigm with a learning and a decentralized decision-making character. The latter is a kind of balance between costs and benefits. The situation of TI measurement can be compared to Foster's (1999: 155) study on justifying knowledge management investments. He proposes to simply estimate the benefits (based on interviews and on average savings of human cost/hour), and to compare these benefits to total implementation and running costs. For example, the CEO of NutraSweet estimates the worth of intelligence activities to be about $50 million a year (Kahaner 1997: 233). In addition to financial results, Foster proposes building the "business case" in a balanced scorecard format. However, the problem is always to estimate the real influence of TI results on decision-making and on positive business performance, and to estimate the indirect influence of TI.

On the other hand, the company can measure the efficiency of the TI process at each stage of the process. There are two ways: benchmarking with competitors and defining indicators of the process. Benchmarking could be conducted indirectly by means of professional organizations such as SCIP (Society of Competitive Intelligence Professionals) or EIRMA (the European Industrial Research Management Association). Quantitative indicators could be, for example, number of sources (journals etc.) or number of conference visit per year, and so forth. These indicators could be compared to the effectiveness of the TI activities at the end of each year. However, there are no studies dealing with such indicators.

There is some literature, mostly in French, on how much companies spend on intelligence. Based on a survey of companies who practice intelligence activities, Ribault et al. (1991: 119) suggest 0.1 to 0.8 percent of sales, but with a median closer to 0.1 percent. They compare this number to the spending in Japanese companies, which is 1.5 percent (in addition to 3 percent expenditures on R&D on average). Hassid et al. (1997: 107) discuss some investment and running costs for SMEs. They are indicated in Table 4.4.

Even if these amounts are indicative, they show that TI costs cannot be neglected. As Sveiby (1997: 153) notes: "R&D is sometimes treated as an investment, sometimes as a cost." The same statement seems to be valid for TI.

The numbers above match more or less with the above-mentioned percentages of Ribault et al. (1991): Based on about 0.1 percent of sales for TI expenditures, this is in line with Hassid et al.'s indications in a 500 employee company of about US$66.5 to 183.75 million turnover, which is quite realistic.

Table 4.4 Some investment and running costs of an SME intelligence systems

Element of expenditure	*SME with 50 employees*	*SME with 500 employees*
Information system (investment costs) [US$]		
Hardware	1,750 to 3,500	8,750 to 43,750
Software	< 1,750	8,750 to 17,500
Time spent and human resources (running costs)		
Top management	3 hours per week in minimum	0.5 days per week in minimum
TI manager	0.5 days per week in minimum	full-time
TI workers	2 hours per week in minimum	2 hours per week in minimum
Information collection (annual running costs) [US$]		
Database connections and internet	3,500 to 8,750	5,250 to 17,500
Field work (conferences etc.)	3,500 to 8,750	17,500 to 43,750
External studies	1,750 to 3,500	8,750 to 35,000
External services	3,500 to 8,750	35,000 to 87,500
Total running costs	12,250 to 29,750 min. 470 hours/year manpower (one sin gle TI worker)	66,500 to 183,750 min. 2300 hours/year manpower (one sin gle TI worker)

Conclusion

Despite the economic importance of small and medium-sized enterprises, there are in general just a few studies conducted on and providing concepts developed for SMEs. How did this happen? On the one hand, there is no generally valid definition of SMEs, and somehow, things one cannot define, one cannot study (accordingly "things you cannot measure, you cannot manage"). On the other hand, research in management seems to give more attention to large firms because problems appear to be more complex than in SMEs. Indeed, problems in large firms are complex, but some of the large firm's problems also occur in small firms (for example fast changing environment), and there are some additional problems typical for SMEs (for example resource restriction). Concerning the common problems, several studies showed that concepts for large firms cannot simply be transferred to SMEs ("a small business is not a little big business!"). Thus, research on small and medium-sized companies in all management research fields is necessary.

This is also true for research in TI. Discussion about diverse aspects in this chapter showed **two interdependent gaps**:

There is a lack of systematic TI approaches in the existing literature on SMEs.

Present research on SMEs is typically about general management issues, more or less along the value chain. Technology issues are discussed in an extra chapter about innovations in the majority of cases. However, the discussion is mostly about advantages and disadvantages of SMEs in relation to large companies in R&D. In fact, recommendations on how to manage these technology issues in SMEs are very rare. Explicit recommendations about the management of TI are altogether lacking. It is also not evident how to gain useful insight for TI from related issues like strategic, technology, innovation and knowledge management because there too, concrete recommendations are rare.

There is a lack of systematic approaches geared to SMEs in the existing literature on TI.

Previous research on TI does not solve SME-specific problems. There is a lot of research about TI methods, which *a priori* could be valid for SMEs. Research on the management of TI focusses mostly explicitly or implicitly on large companies. However, discussion in this chapter showed that existing concepts (for large firms) are not practical for SMEs, and proposed methods are less applicable in SMEs because of their resource restriction. Thus, SMEs are gray spaces in TI research. These gaps should be closed, which is the aim of this book.

5
The Research Framework

Gaps of Technology Intelligence in technology-based SMEs

The previous chapters showed a lack of Technology Intelligence (TI) in technology-based SMEs from a theoretical and a practical point of view. The practitioner's voice demands a solution that meets SMEs' specifications, of which resource restriction is a primary concern. Resource restriction seems to be the principal reason for numerous cases in SME-specific research, where solutions for large companies are very often scaled down in order to be appropriate for SMEs. However, the literature is rather suspicious of such transformations. A look into the reality of TI in technology-based SMEs showed that the topic is of concern in practice. Even though companies generally do some activities in order to be informed about what happens in their technological environment, they rarely do it in a coordinated manner. This implies three risks. Firstly, there is no overview of activities, which means that it is not clear who has what kind of information, and if there are redundancies. Secondly, there are probably some observation areas that are not covered but should be. Thirdly, activities do not adequately harmonize with the company's strategy. Thus, there is a need in practice to give TI activities a system.

To summarize, the most important gaps and insights from the literature survey and from reality at a first look into the field of Technology Intelligence in technology-based small and medium-sized companies are as follows:

- There is a call from reality to identify, collect, analyze, and apply relevant information about the company's technological environment in order to improve the quality of decision-making (= Technology Intelligence).

- The SME literature does not provide solutions for Technology Intelligence. In fact, other contributions about SMEs additionally identified some problems which are relevant to the field of TI, for example corporate strategic planning, resource allocation in R&D, and communication in R&D.
- The general Technology Intelligence literature does not provide solutions for SMEs. In fact, most solutions are designed for large companies. These solutions are usually unsuitable for SMEs because of the latters' resource restrictions.
- However, some SMEs already do intelligence activities, but they are neither coordinated nor planned. Most companies wish to give the activities a more systematic order to cover more or less the entire value creation process of TI activities.

Empirical research design

Since there are too many gaps in this topic, an empirical research study conducted in Switzerland during a period of three years forms the basis for this book. Because of the insufficiencies mentioned above, an empirical and explorative research design was to be preferred initially. Most current empirical and explorative studies are based on case studies. The choice of the case-study research can be explained by Kubicek (1975: 61). He argues that case studies are best for the very early stages in research of an organizational problem. They need relatively little effort and bring plenty of suggestions for further research on this topic. However, there are some weaknesses of case studies, which are mainly that they provide an inadequate generalization of insight.

With regard to this concern, a two-stage research procedure has been chosen for this book: specific insight resulting from one action research case is analyzed in a second stage in the context of other companies. In this sense, the action research case serves as an inductive generation of insight into a still little explored field of research in TI, which shall be strengthened and validated with case studies in a second part. The underlying procedure follows the principles of the multiple case-study design of Yin (1988: 53): "Every case should serve a specific purpose within the overall scope of inquiry." Nevertheless, this book does not claim an overall validation of the learning from the action research case. This would not be possible because of two facts: the sample of case studies is always restricted by the number of companies studied, and then, the insight gained from a case study always reflects one company's specific reality. Thus, the aim of this research is to explore how an SME can design a TI system, and then to present a set of management principles to guide SMEs to an appropriate solution for TI.

The dominant significance of the action research case is the result of close research cooperation between the author and the company. This case should build, in an explorative manner, the basis for the discussion of the elements of a Technology Intelligence system for SMEs. The action research is described in detail below. In accordance with the two-stage research design, a detailed description of the validation cases will follow.

Action research

The term "action research" is attributed to Lewin (1946). His work seems to be fundamental to the modern understanding of action research: "He created a new role for researchers and redefined criteria for judging the quality of an inquiry process. Lewin shifted the researcher's role from being a distant observer to involvement in concrete problem solving" (Greenwood & Levin 1998: 19). Since the 1970s, Kubicek (1975) has observed an intensified attention to action research – he also calls it "research by development" – in organizational research. He designates action research as an approach, in which practitioners and scientists jointly plan and implement new organizational concepts. Moreover, the involved scientists try in turn to systematize and generalize their experiences. Thus, action research is action oriented. This means that the researcher is able to actively influence the research object, which in contrast is not possible with passive approaches like pure case studies. In case studies, the researcher is limited to formulating questions and interpreting the empirical results.

Three central terms in action research are:

- *Research* (knowledge generation),
- *Participation* (the participatory process in which everyone involved takes some responsibility),
- *Action* (jointly elaborated options of action).

One can only speak about action research in its proper sense if all of these aspects are considered in the study. To clarify the content of action research, two current definitions are given below, and then illustrated in Figure 5.1:

- *Greenwood & Levin (1998: 4):* "Together, the professional researcher and the stakeholders define the problems to be examined, co-generate relevant knowledge about them, learn and execute social research techniques, take action, and interpret the results of actions based on what they have learned."
- *Cunningham (1993: 4):* "Action Research is a term for describing a spectrum of activities that focus on research, planning, theorizing, learning, and development. It describes a continuous process of research and learning in the researcher's long-term relationship with a problem."

In fact, the "influencing" aspect of action research makes exactly the difference between action research and traditional case-study research, where taking action would be considered as biasing the environment under study.

Facing the often-cited contrast of qualitative and quantitative research methods, action research adopts a rather neutral position. In principle, action research allows any kind of method of social science. "Surveys, statistical analysis, interviews, focus groups, ethnographies, and life histories are all acceptable, if the reason for deploying them has been agreed on by the action research collaborators and if they are used in a way that does not oppress the participants" (Greenwood & Levin 1998: 7). Thus, action research seems to be very promising for explorative studies in organizations. Both, the research community and the organizations benefit from the experience gained during common design and implementation of new concepts. The situation of action research and how the scientific and industrial community can benefit is presented in Figure 5.2.

The great opportunity of this kind of research lies in the two supplementary dimensions of action and participation. Therefore the action research project in this book is based on a cooperative problem solving and learning approach. It is a tightrope walk of interests for scientists and practitioners. This is particularly challenging to the management of the conducted action research project.

In practice, action research is normally restricted to one single company because of the limited degree of readiness of organizations to cooperate (Kubicek 1975: 71). Anyhow, the restriction to one single company requires further comparative studies in other companies to deepen the understanding. To accommodate this purpose, a two-stage research approach has been chosen. The second stage is about the further validation cases, which are described in the following.

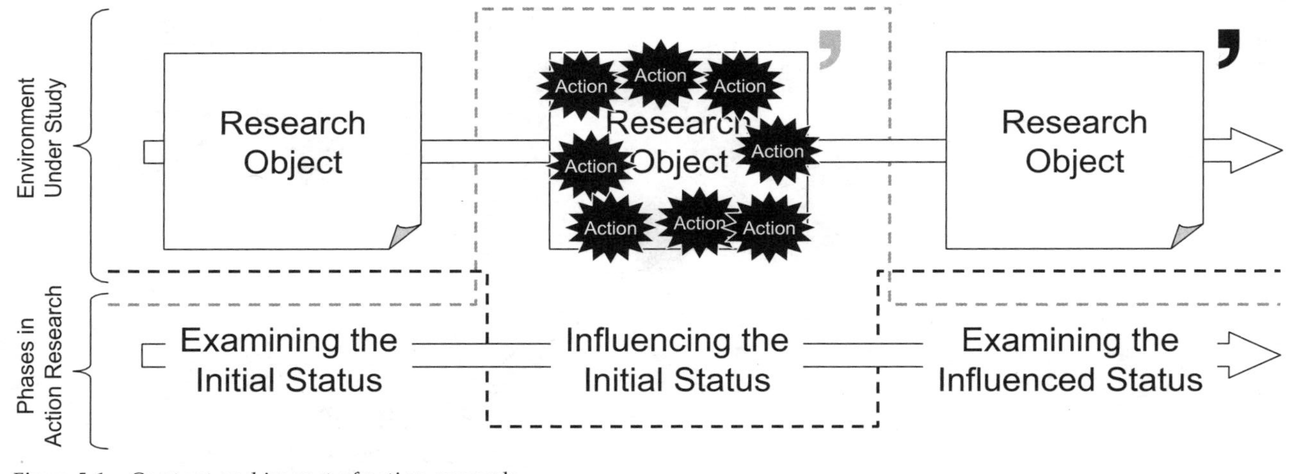

Figure 5.1 Content and impact of action research

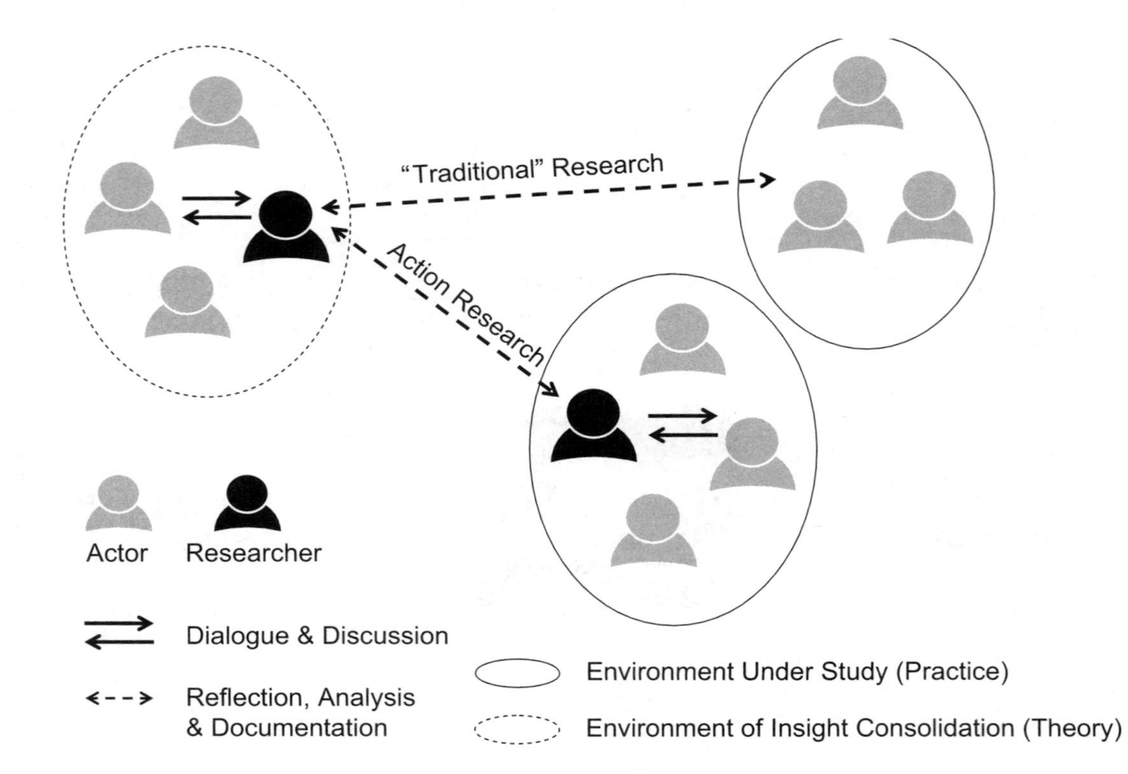

Figure 5.2 Benefits to scientific and industrial communities from action research

Further validation cases

The complementary validation cases serve to discuss and deepen the understanding of the previously generated elements of a TI system in an SME. These generated elements can be understood as "theory." It seems that case studies are a more promising method of validation than quantitative testing: "Any finding or conclusion in a case study is supported by a chain of multiple evidence from different sources, and is therefore more convincing and accurate than a finding or conclusion from a survey" (Chetty 1996: 77). For this validation, two types of case studies have been chosen. The aim of the first type is to "test" and explore the previously built theory in a context that is similar to the one used for the action research environment, while the second type emphasizes comparing insight in another context. This implies that there is still some action in the cases of the first type, while cases of the second type are descriptive.

Even though it is not possible to make an overall generalization of the generated elements in action research through adoption of this multiple case-study approach, nor to build a universally valid theory, the opportunity to cope with different realities, i.e. different contexts of the companies, makes them very useful in the almost unexplored field of Technology Intelligence for SMEs.

6
Action Research at Institute Straumann AG

According to Figure 5.1, a case study initially examines the original status of the research object, which is a technology-based SME. This chapter is devoted to knowledge creation by influencing the research object, for example action research. The conclusion will discuss the influenced study environment.

Case study: "Straumann 1999"

The information in this chapter was collected through interviews with employees and document analysis in October/November 1999 at a Swiss medium-sized, technology-based company: Institute Straumann AG (hereafter: Straumann). However, in order to increase the understanding of the company's context, more current information, for example about Straumann's financial situation and market data, will complete this case study.

The company's characteristics and history

In 1999, about 470 (2000: 540; 2001: 670) employees contributed to net sales of US$88.3 million (2000: 112.1 million, 2001: 136.9 million). The company is headquartered in Waldenburg (Switzerland) and has subsidiaries in Europe and overseas. R&D is pursued entirely in Waldenburg, and products are manufactured in Villeret, a village in the Swiss watch industry valley with its first-rate international reputation for incontestable quality. In 1954, Prof. Dr. H. C. Reinhard Straumann founded a private research company known as Institute Straumann AG in Waldenburg near Basel, which specialized in metallurgy and physics. The alloys developed by the Institute Straumann for springs in mechanical clock and watch movements established its excellent reputation and are still used today by many of Switzerland's best-known watch manufacturers.

During the 1960s, the company began using its metallurgical expertise to develop metallic implants for the treatment of bone fractures. A collaboration with the AO/ASIF (Arbeitsgemeinschaft für Osseosynthesefragen, Association for the Study of Internal Fixation), a professional association in the field of internal bone fixations, was set up. Subsequently, Dr. H.C. Fritz Straumann, the founder's son, concentrated on research on ultra-fatigue-proof implant alloys and tissue tolerance. By the late 1980s, the company, which by then was active on an international scale, had successfully established itself as a manufacturer of implants and instruments for internal fixation and oral implantology. Meanwhile, Straumann and some leading academics in dentistry have created the "International Team for Oral Implantology" (ITI). The ITI is a scientific alliance of leading scientists and practitioners in the field of oral and extra-oral implantology, working on an unpaid and altruistic basis. Today, ITI consists of more than 200 members from around the globe. The team is well-known and has the reputation of being an honorable, serious and professionally competent group. The scientific accomplishments and established products of the ITI Implant System, manufactured after intensive development and testing, are successfully employed worldwide. The System fundamentals and some of the principles have become standards in implant dentistry. The main sponsor of the ITI is Straumann. The corresponding countervalue is exclusive license rights on ITI patents.

Since 1998, Straumann has been a fast growing public company. Nevertheless, the Straumann family continues to be major shareholders. (See Figure 6.1.)

Products and market

Straumann develops, produces, and markets implants, prosthetic parts, and instruments for oral and extra-oral implantology and craniomaxillofacial internal fixation world-wide. These products and services enable professionals in dentistry and oral surgery to treat patients successfully, and to improve their quality of life.

A short technical description of the products

The company's main product line is the ITI DENTAL IMPLANT SYSTEM. Complementary product lines are the Straumann Orthosystem and the oral and maxillofacial surgery. The latter was developed by Straumann R&D in partnership with Medartis AG.

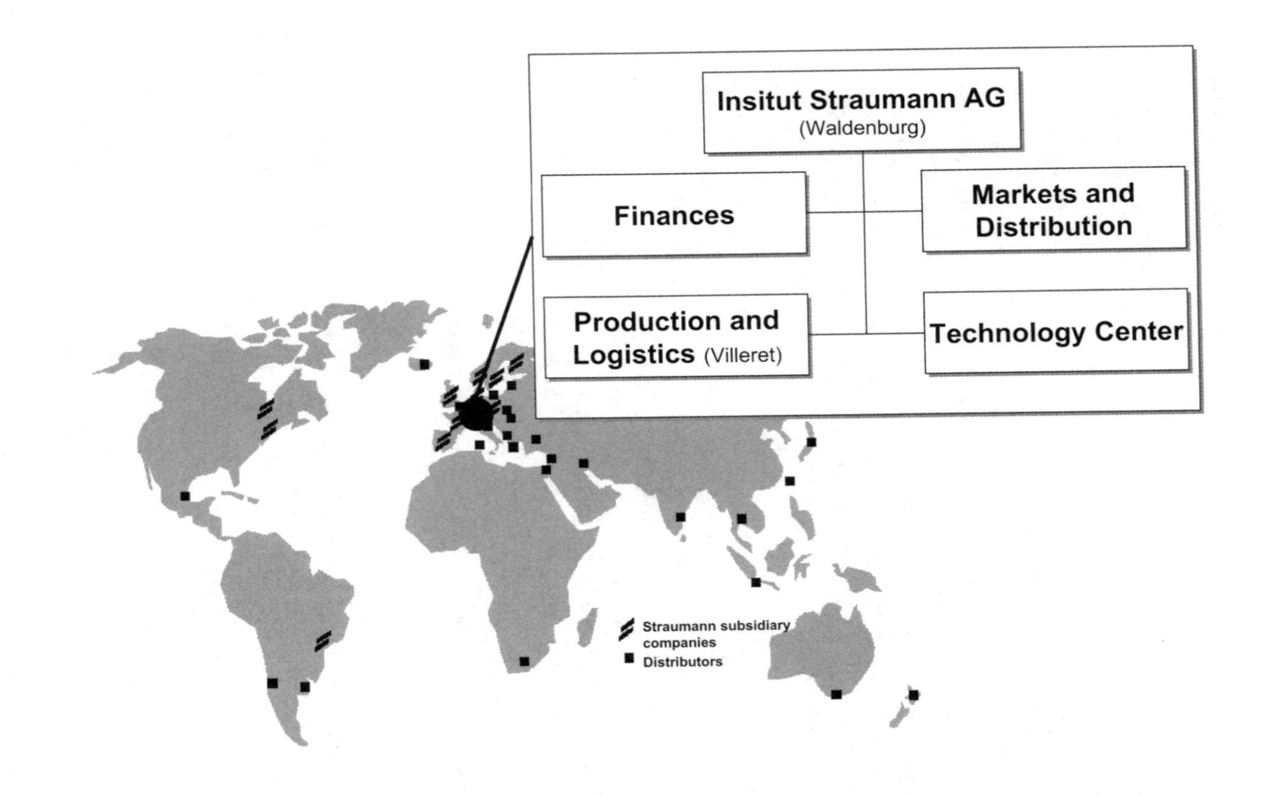

Figure 6.1 Straumann in Switzerland and subsidiaries in the rest of the world

Because this product line was transferred entirely to Medartis in 2000, the case study describes primarily activities concerning the first two product lines.

Dental implants are titanium screws or cylinders which are implanted into the upper or lower jawbone. They provide an artificial root for single tooth crowns, bridges or overdentures in the case of a partially, or completely edentulous jawbone. The connection between the basic implant and the crown/bridge is provided by the suprastructure or abutment. When compared to conventional restoration with crowns or bridges in the case of single or several missing teeth the advantage is the fact that no neighboring healthy teeth have to be charged in order to fix the crowns/bridges. Furthermore, the functional load on the jawbone in the area of the implant is maintained so that no atrophy or pathologic reduction in bone mass occurs. The disadvantage of the more elegant and physiological use of implants is the higher price and the fact that implants are almost never covered by insurance.

Some special features distinguish the ITI system from the competitors and make Straumann leaders in this technology area:

- *SLA surface*: This sand-blasted, large grit, acid-etched (SLA) surface reduces healing time from three months down to six weeks. This allows earlier loading of the implant. The sand-blasting creates a macro-roughness which gives a superior stability in the bone. A micro-roughness is derived from the acid-etching to increase cell activity on the surface in order to increase stability.
- *Minimally invasive*: The ITI system utilizes a nonsubmerged, single stage method for the following reasons: firstly, the single-stage surgery procedure limits patient trauma and reduces costs by shortening the duration of treatment. Secondly, there is no microgap between two matting parts, a potential site for plaque retention, in the tissues. Thirdly, the gingival seal is formed during the primary healing phase. This "philosophy" is more and more imitated by competitors.
- *Anatomically shaped neck section*: Aesthetic results at the crucial junction with the soft tissue extending to above the gum.
- *Cemented suprastructure*: The cemented restoration is an excellent alternative to the screw-retained approach. It reduces costs significantly, normally requires less maintenance, and meets esthetic concerns. On the one hand, once the prosthesis is cemented, it cannot be simply removed. On the other hand, since an implant replaces a natural

tooth, the cemented solution does not have the inconveniences associated with natural teeth. (See Figure 6.2.)

Market dynamics

There is a two-stage customer logic in Straumann's business: the direct customers are private dentists and clinics, the patients are indirect customers. Although Straumann started a public awareness campaign in some US states, the product choice is normally made by the professionals. Therefore, a trusting relation with these professionals is crucial for success in Straumann's business.

The dental implant manufacturer contributes about 10 to 15% to the final product value (surgeon: 65–70%, dental laboratory: 15–20%). To give an example, for a single tooth replacement: Traditional dentistry (3-unit bridge) costs US$2,650, an implant-supported restoration costs US$1,830. (Average indication for treatment in Switzerland. Cost may differ considerably depending on individual clinical condition.)

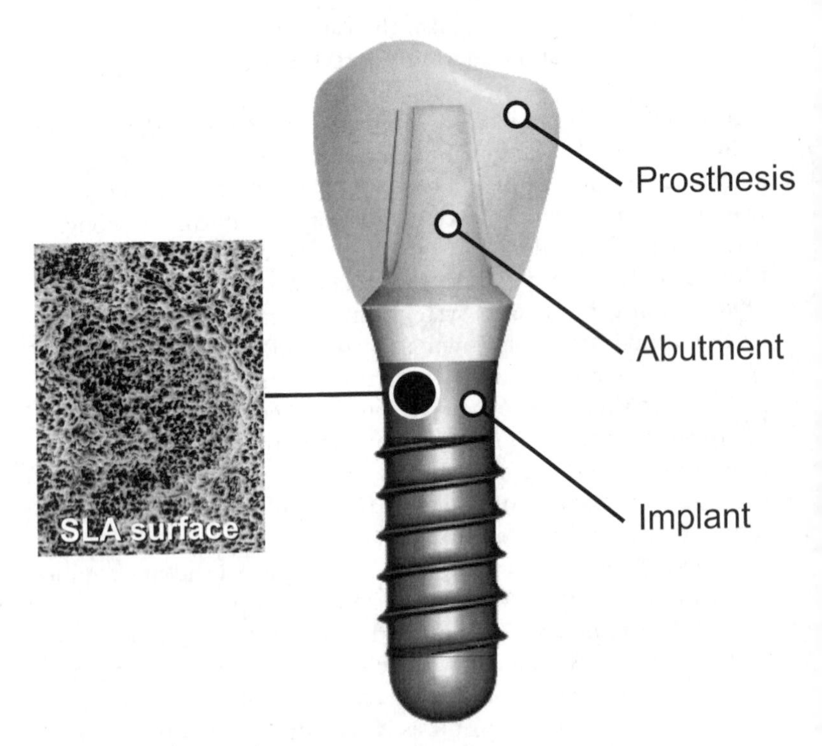

Figure 6.2 Dental implant from the ITI system

Implant treatment in general is not reimbursed by insurance with a few exceptions in Sweden, the Netherlands, and Germany, where treatment is partially reimbursed.

Dental implants are mostly used in industrialized countries. Today, from about 40% of population with incomplete dentition in industrialized countries, just 0.4% are treated with implants. This makes an actual market volume of about 51 million people, compared to a market potential of about 320 million people. However, a 15% growth rate per year increases the market penetration considerably. (See Figure 6.3.)

Competitors

The market shares of the most important competitors are depicted in Figure 6.4.

A market consolidation took place over 1999/2000. Nobel Biocare strengthened their leading position through the acquisition of Steri-Oss, Calcitek (a Sulzer Medica subsidiary) purchased Paragon, and a new entrant to the market, Dentsply, acquired Friadent.

In this competitive environment, Straumann is the fastest growing player. Its growth rate of about 35.4% in 1999 (27.1% in 2000, 21.8% in 2001) clearly beats the average market growth rate of 15%.

An overview differentiating the products of the four big players is given in Table 6.1.

R&D at Straumann

Straumann products emerge from interdisciplinary work between dentists, engineers, physicians, and biologists. Basic research is conducted

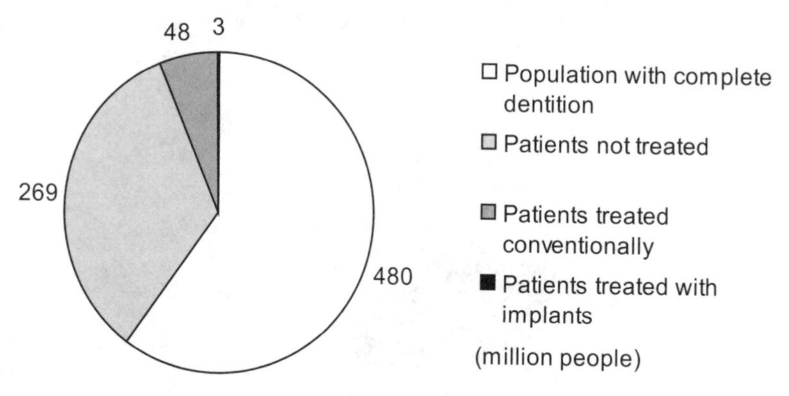

Figure 6.3 Market penetration in industrialized countries

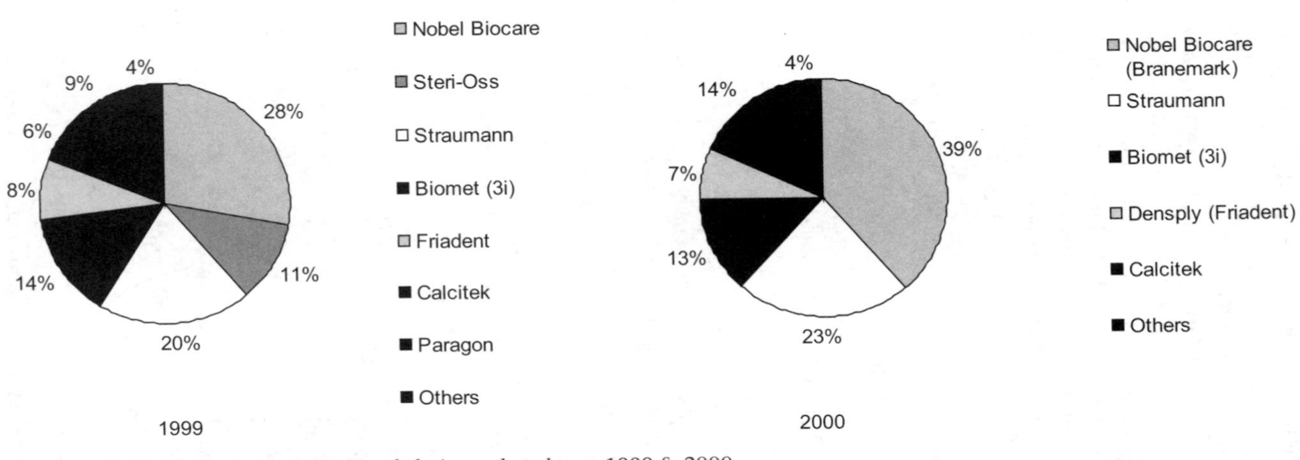

Figure 6.4 Important competitors and their market shares 1999 & 2000

Table 6.1 Overview of Straumann and competitors

	Straumann	*Nobel Biocare*	*Biomet*	*Calcitek*
Brands	ITI MODUS	Branemark System STERI-OSS PROCERA	3i OSSEOTITE	CALCITEK PARAGON
Market share (1999)	US: 11% EC: 25%	US: 39% EC: 32%	US: 27% EC: 11%	US: 15% EC: 6%
Sales	US$89 million	US$175 million	US$67 million	US$69 million
Differences	One stage Rough surfaces 6–8 weeks healing time Immediate loading possible Morse-taper	Two- and one-stage Smooth and rough surfaces 3–12 month healing time Immediate loading possible External hex and internal connection	Two- and one-stage Smooth and rough surfaces 8 week healing time Immediate loading possible External hex and internal connection	Two- and one-stage Smooth and rough surfaces 3 month healing time Immediate loading possible External hex and internal connection

in national and international collaboration with universities and the ITI. Applied research and development are done internally and/or through cooperation projects with specialists, for example with dentists. Because of Straumann's worldwide collaboration projects in basic research, it has a decentralized character. Applied research and development are centralized in the headquarter site in Waldenburg.

In 1999, Straumann spent about 8–10% of net sales for R&D. In terms of head count, there were 46 employees directly involved in R&D.

R&D mission

The R&D mission at Straumann is:

> ... *to keep, increase, and transform the innovation power of Straumann into market-oriented and profitable product concepts.*

The "innovation power" can be understood as technology and innovation potential, which is to be managed. In fact, in order to stimulate market orientation, product marketing is very close to R&D at Straumann. To "increase the innovation power" is the commitment to a proactive technology leadership.

R&D structures and tasks

From 2000 on, research, development and product marketing have been concentrated in the so-called "Technology Center." Previous to 2000, research was separated from development organizationally and geographically in another building. With the new Technology Center, communication, and therefore innovation, should be accelerated.

In order to accomplish the R&D mission, the structure of the Technology Center was chosen (see Figure 6.5).

The head of the Technology Center is a representative to the executive board. He is responsible for and coordinates the tasks within the Technology Center. There are four business lines and four staff functions at his disposal.

Research focuses on applied research in the fields of metallography and biochemistry. Its main tasks are to understand materials and their interaction with bone cells, to develop appropriate surfaces, and to test materials and components. New product development and rapid deployment are tasks of **Engineering & Development**. In addition to rather classic marketing and communication tasks, incremental product changes are conducted by **Product Marketing**, with technical support from the engineering group. The **Service Centers** group supports disciplines such as development mechanics, regulatory affairs, quality techniques, and product services.

Since **Clinical Research** tests and documents new products throughout the development process, it is considered a cross-functional discipline to research, development and marketing and figures therefore as staff unit. The **Technology Transfer** staff function represents an internal technical consultation, mainly for patent concerns. The **Screening** function (not detailed yet) deals with idea evaluation for product development. **Patents** are administered in the fourth staff function.

As mentioned before, R&D is pursued in close collaboration with the ITI. Straumann works directly with and is represented in three committees for development, training & education and research, which in turn works in close collaboration with universities and research institutes. The formal relationship between Straumann and the ITI is shown in Figure 6.6.

The financing of research is the central activity of the **ITI Research Committee**. It supports research projects for implantology and its

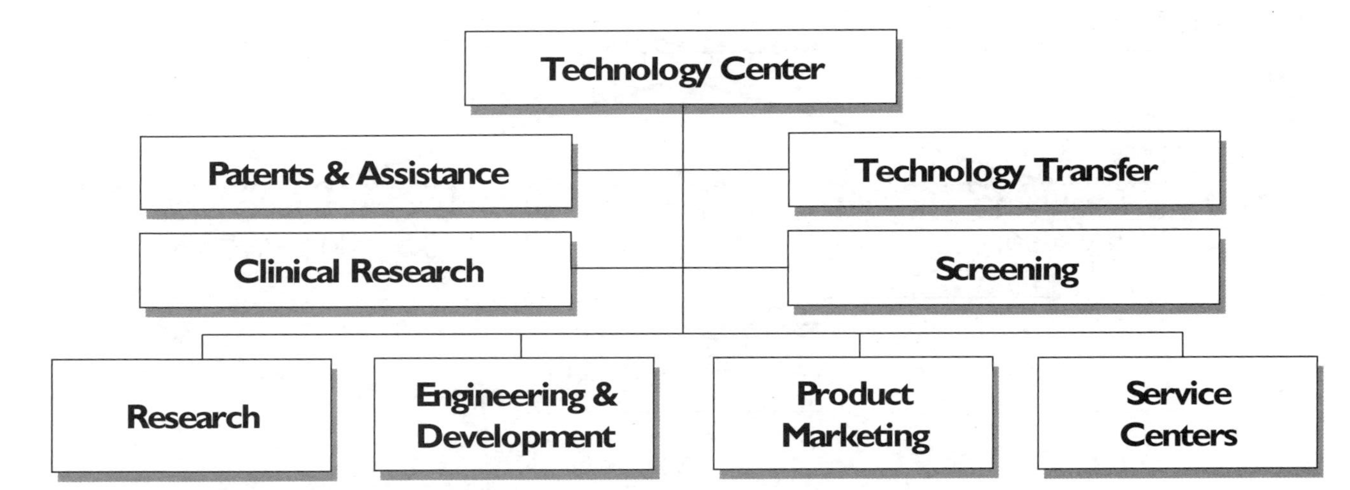

Figure 6.5 Organization chart of the Straumann Technology Center

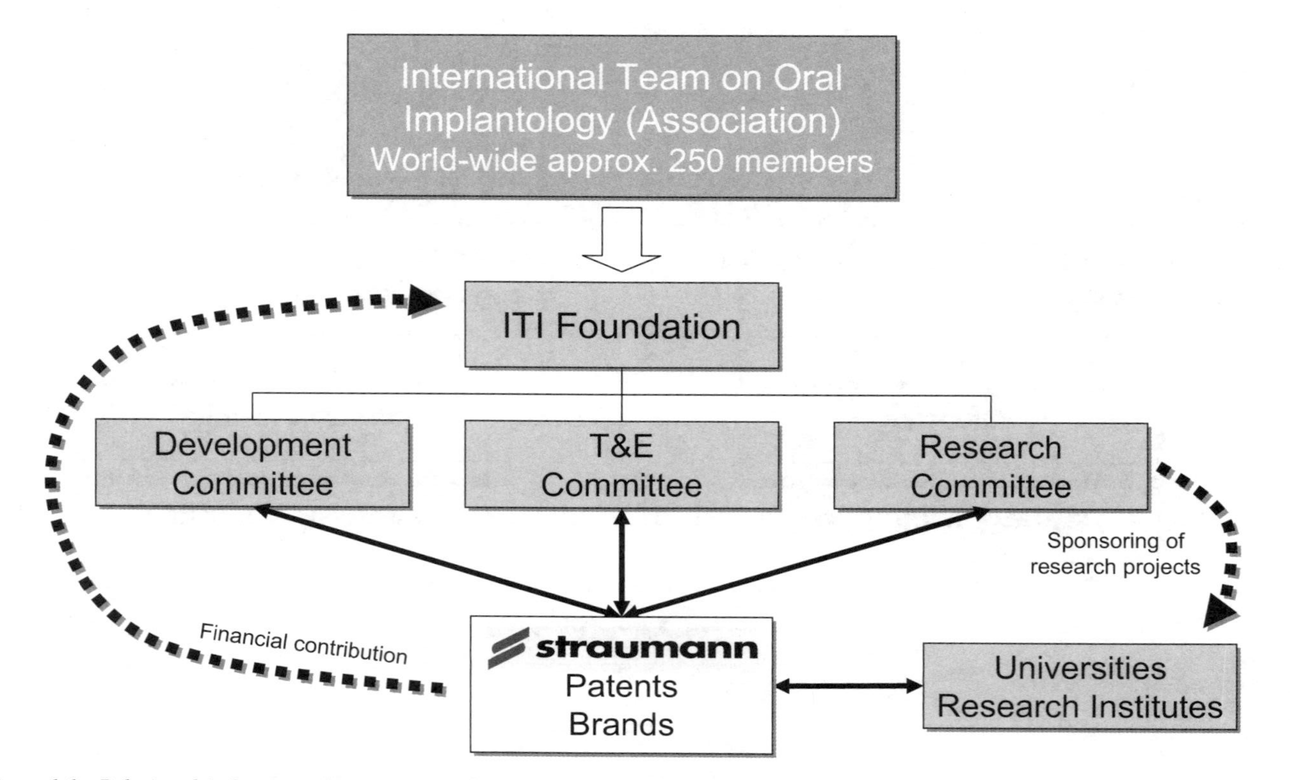

Figure 6.6 Relationship between Straumann and the ITI

allied interests, and today is counted among the most important institutions in the Swiss medical research field. Interested people can submit applications for financial support of research projects. The committee meets twice a year to decide on project support. Thus, since its establishment in 1988 up to the year 2000, more than US$8 million have been distributed for the support of research projects. The majority of these funds were allocated to people who were not members of the ITI.

The **ITI Development Committee** unites the requirements of the practitioner with those of the researcher. It introduces ideas and findings from clinical experience, provides fresh impulses for both new development and further development to the experts of Straumann and carries out clinical experiments.

The **ITI Education and Training Committee** conveys the basic knowledge as well as the latest findings from research and development. It coordinates internal ITI meetings and organizes congresses and symposia. Since 1988, an ITI World Symposium has been held every other year. The venue alternates between Europe and North America. The ITI World Symposium has become one of the most important events in the field of implant dentistry. At the last three events held in 1996 in Basel (Switzerland), in 1998 in Boston (USA), and in 2000 in Lucerne (Switzerland), up to 2,000 participants attended each symposium. Collaboration with universities and research institutes (for example with the University of Texas Health Science Center at San Antonio and the School of Dental Medicine, University of Berne) are both, by means of the ITI and through direct contact.

The R&D process

Product and technology development is organized by projects. Project duration ranges from a few months to several years. This can be explained by the natures of the different projects. For example a project about further development of the surface has rather a permanent character and, thus, takes a longer time than a simple product upgrade.

Straumann does not explicitly differentiate between incremental and radical innovation and there is therefore no proportional resource allocation to these two project types. Anyhow, there is an implicit difference in terms of project handling. The more the project is "incremental," the more detailed the project management is set up. In turn, the more the project is "radical," the more vague and "open-minded" the project goals are. An estimation of allocated resources to both project natures would be about 80% incremental and 20% radical.

Generally, the impulse for a R&D project emerges from a market need. There is very strong dominance of market opinion through the ITI. ITI members are considered to be opinion leaders in the oral implantology market. In fact, through the network within the ITI, especially in the different committees, some ideas "just come up," and are then developed. There are typically vague ideas from single ITI members, who ask Straumann for collaboration in development of the idea. There is a constant interaction between the collaboration partners during a development project. However, a formalized process of this collaboration does not exist. In addition to ideas emerging from ITI collaboration, ideas can rise from other external and internal sources.

The forthcoming steps in the innovation process are shown in Figure 6.7. In fact, this is just a representative figure. The formalized process begins with the functional specification. Very important is the product launch process being in parallel. Clinical studies, or field test are necessary to prove safety and effectiveness of new products. They take a long time sometimes and often determine the critical path of a product development process.

R&D tools (methods and infrastructure)

R&D is strongly supported by methods and an adequate infrastructure. Examples of technical methods are the Finite Element Method (FEM) and computer-aided methods (CAx: for example CAD, CAM, and CAE). However, some are outsourced. Straumann also utilizes creative methods such as bionics, and projects are conducted with value analysis. Modern technology management methods, for example Technology Portfolios, Technology Value Analysis, Technology Roadmapping, are hardly ever used. Nevertheless, members of the Technology Center wish to utilize such methods more often.

R&D infrastructure is very modern. In addition to the above mentioned technical methods, high-end hardware, like for example a scanning electron microscope (SEM), is available in house. For communication and database purposes, Straumann utilizes Lotus Notes.

Technology intelligence at Straumann

Straumann does not run a systematized Technology Intelligence (TI) system or process yet. However, like other interviewed companies, Straumann implicitly pursues some TI activities. In fact, during this

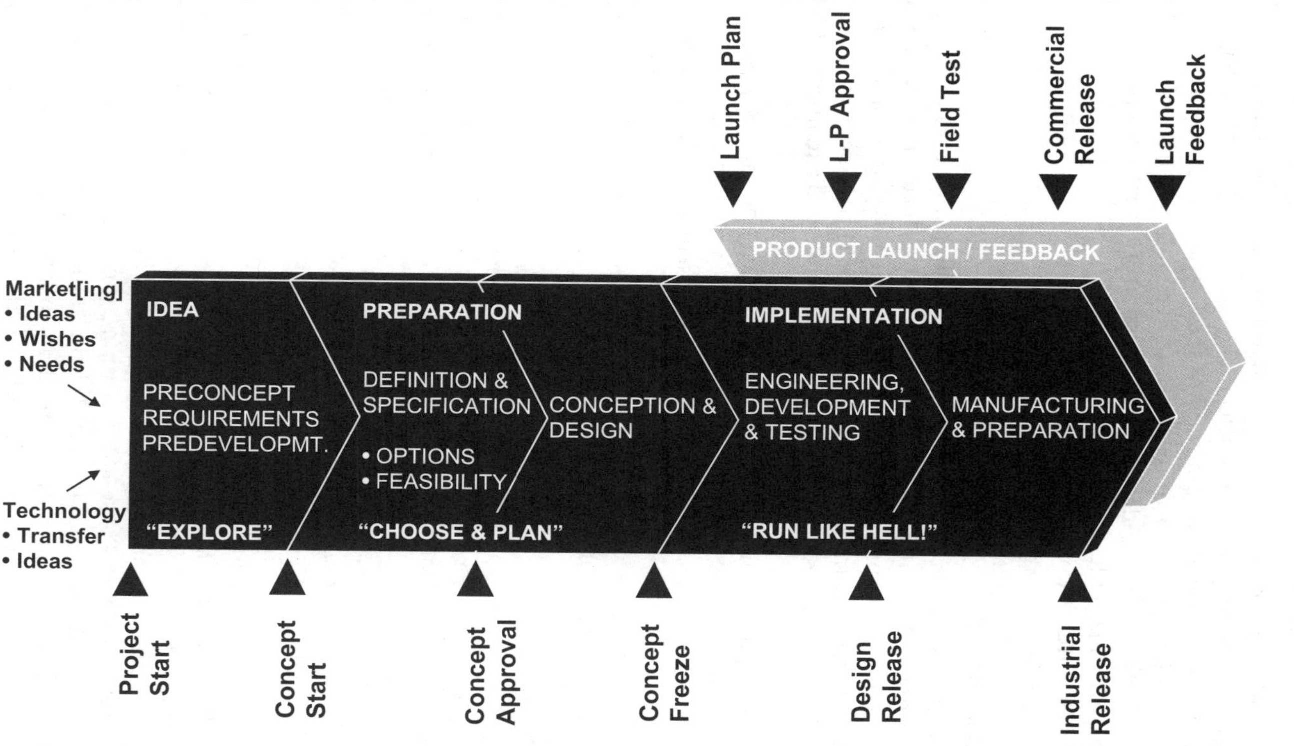

Figure 6.7 The innovation process at Straumann

case study a process called the "screening process" was planned, but has not yet been realized. The screening process was developed to give the "fuzzy front-end" of innovations more structure. The content of the screening process is of relevance to Technology Intelligence and will therefore be included in the discussion. However, no practical experience may enter the discussion.

Any identified TI elements are discussed hereafter by means of the TI value chain.

TI management

TI activities are not formally managed at Straumann, i.e. one cannot speak about designing, directing, and developing TI activities. Because TI activities just "happen" in an uncoordinated way, it is difficult to guess the relationship between them. Therefore TI activities seem to be almost individual initiatives, rather than a coordinated and institutionalized system.

An exception is the "screening process," which is in the design stage. Unlike the idea of a synchronous-parallel implementation to the design process, Straumann decided on a sequential implementation. The reason for this was the fact that a process-owner (screening coordinator) had not yet been defined. The company intended to hire a person especially for this job.

TI mission and goals

TI mission and goals should define the purpose of TI activities. Since there is no TI system, there is no clear TI mission that copes with business mission and strategy. However, the screening process has a main goal, which is:

> *to lead to effective and efficient decision-making before product development.*

By this, McDonald and Richardson's (1997:124) view is fulfilled partly "to support business decisions." The screening process's focus is limited to product development decisions. In fact, the screening process evaluates any idea to decide whether a development project should be started. These ideas are more short-term and well developed rather than oriented toward the future. Thus, the screening process's scope could be assigned to the "keeping abreast" area. Influencing factors and contents of the screening process are represented in Figure 6.8.

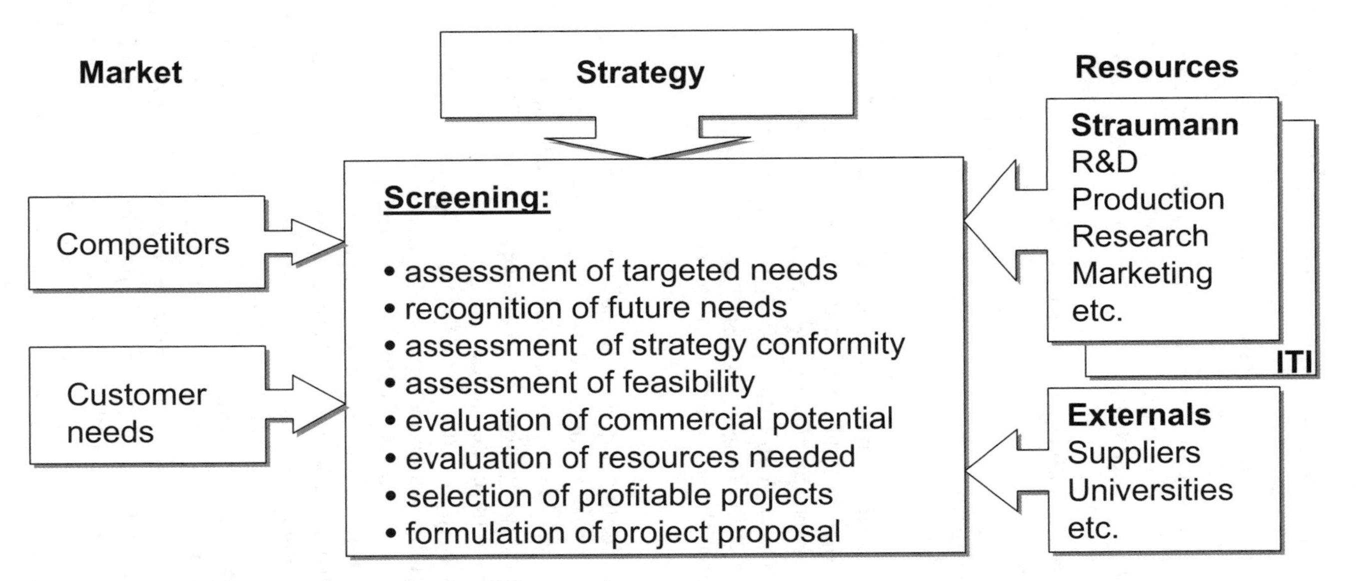

Figure 6.8 Content and influencing factors of the screening process

The screening process does not support decisions or concerns other than product development. Long-term and strategic decisions especially do not have a firm information base.

The focus of Straumann's business activities is mainly on oral implantology. Thus, unsystematic and informal "everyday" TI activities (besides the screening process) are supposed to focus on oral implantology or related fields. This is true for most activities. However, there are some TI activities that look beyond this field. In the majority of cases such initiatives are kind of exploration projects that are induced *ad hoc* by the top management in order to assess a technology that could be of interest to Straumann. But these projects do not have an institutionalized status.

TI structures

Even though Straumann does not pursue a formalized TI system, some people and units should be discussed with regard to TI activities. These "elements" are staff functions (screening, patents, technology transfer), line functions (research, clinical research, engineering & development, product marketing, standard marketing), top management and external structures (ITI, lead suppliers and users).

Screening

Since the screening function is the nearest activity to Technology Intelligence, it shall be discussed first. During the interviews, this function was not yet in operation. However, some reflections about who should participate in the screening process, and how involved people should interact, have been made.

Screening is supposed to be an interdisciplinary function, and is therefore run by various people from various domains. The design of the screening function is envisioned as having four permanent screening team members: one screening manager and three people from research, development, and marketing. Other people should meet the team occasionally to provide expertise. The permanent core team should attend a regular and institutionalized meeting. As Figure 6.8 indicates, the ITI and other external institutions should also be involved for expertise. Thus, the head count for the screening process is not negligible. This lets us conclude that the screening process is considered an important management process. The screening manager is a staff function that is attached to the Technology Center (see Figure 6.5). Thus, screening is a technical function. However, an overall business view should be guaranteed by a direct link to the head of the

Technology Center, who in turn is a member of the management board.

Patents

The literature states that patents are an important source of intelligence. Straumann has a patent staff function that is also attached to the Technology Center. Straumann is subscribed to the most important patent classes for oral implantology. A package of new patents is rotated among several R&D employees in order to check whether a new patent infringes on any Straumann patents and to make sure that Straumann developments do not become critical regarding the latest patents. The opportunity to be inspired by new developments in oral implantology is particularly a concern of Technology Intelligence. However, the inspection of patents remains voluntary among employees, and reality shows that this task is barely assured. Furthermore, just a few employees know where to communicate insight from patents, making patents almost unutilized for TI at Straumann.

Technology transfer

Another staff function of the Technology Center is called technology transfer. This name does not match the fulfilled task of the person in this staff position. In fact, this person has some special tasks, and some of them are of strategic concern. This person is a real "gatekeeper" (Allen 1986: 144). He has numerous contacts to external experts, and he transfers knowledge from these contacts to the company. Thus, he pursues (informal and unstructured) TI activities. However, he is not perceived as a "formal" intelligence source by top management.

Research, clinical research, engineering, and development

Like the technology transfer person, many scientists and engineers pursue intelligence activities. They attend conferences, seminars, and tradeshows, they meet experts (for example, in universities), they meet suppliers, they work in cooperation with others, they read journals and books, and more. However, some problems are that they are not consciously aware of pursuing TI activities, that they are not trained for this (and thus work less effectively and less efficiently), and that they only rarely communicate insight. It has to be noted that all of these TI activities are informal, i.e. they do not follow any kind of TI mission and goal.

Anyhow, a common platform exists in project teams. For any project, scientists and engineers should become informed about facts and trends that are tangential to their project. However, there are no guidelines, nor is this a formalized project requirement. At least communication within a project team is favorable to information exchange.

Products and the "standard marketing group"

Product marketing is attached to the Technology Center. Some of the product marketer's tasks are product placement and customer contacts. This proximity to customers provides very useful insight into technological facts and trends. Therefore, they are the "market correspondents" to the scientists and engineers. However, their knowledge most often just influences product adaptation rather than real innovations. Existing structures barely allow the transfer of this knowledge to other Technology Center employees.

The "standard marketing group" is attached to the market and distribution unit. One task is market research or market intelligence. In fact, the content of market intelligence at Straumann does not cope at all with the definition of Technology Intelligence adopted in this volume. Market intelligence first of all explores market volume development and similar tasks. It seems that one parallel to TI activities could be some analysis methods. However, there is no interaction between this marketing group and the Technology Center.

Top management

Top management seems to be an important element of Technology Intelligence. Firstly, board members and chief executives have numerous contacts with experts in the field of oral implantology, which are important intelligence sources. A particularly fruitful source is the link to the ITI. Secondly, some top managers have a technical background, which allows them to analyze the expert opinions quickly and competently. In addition to this role of "intelligence worker," top management is simultaneously a key intelligence user.

The ITI

The ITI is an external expert network and independent research fund at the same time. As previously described, the ITI unifies about 200 leading experts who are opinion leaders in the field of oral implantology. These may be university professors, dentists, orthodontists, etc. Within various committees they meet several times a year to share experiences, to discuss trends, and to plan projects. In addition to these institutionalized meetings, members of the ITI meet informally at conferences and during their daily work.

The ITI sponsors clinical and laboratory research projects in the field of oral implantology. Funding is legally independent from Straumann. Four different grant programs are available:

- **Small grant application:** Such grants are distributed to research proposals for building up a research group. Maximum sum: US$18,000.
- **Research proposal for clinical or laboratory research:** Such grants are awarded to researchers or research groups who have established a reputation for credibility and thoroughness in the field by establishing a continuous publication record in peer-reviewed journals. Maximum sum: US$90,000.
- **Research program projects:** Such grant proposals may contain several single research projects or several single aspects within a major line of research and will usually be granted for a number of years (3–5 years). These projects are only given to well-established research teams of international reputation. Maximum sum: US$90,000.
- **Single laboratory support:** Such grants are allocated to established laboratories with expertise in a field of interest to the ITI. The support should not exceed the salary of a laboratory technician and the duration of two years.

Technical knowledge, as well as knowledge about future trends gained from these exploratory research projects, is available to Straumann. Thus, these projects are an important intelligence element.

The role of Straumann in the ITI is twofold. Firstly, Straumann is the main sponsor of the ITI foundation. Secondly, some Straumann employees, especially members of the board and chief executives, are also member of the ITI. In this way, a direct and expedient transfer of gained intelligence is guaranteed. At the same time it is up to Straumann to promote explorative studies in a particular field of interest.

Lead suppliers and lead users

Most Straumann suppliers mainly supply raw materials. Experience has shown that new materials seldom emerge from these suppliers but rather from other institutions. Therefore one cannot speak about lead suppliers in its proper definition. In turn, many ideas and propositions emerge from users. Some of the most important users (lead users) are in constant contact with Straumann through the ITI and product marketing. However, Straumann management is attentive to one point regarding lead users: their thoughts and ideas are mostly restricted to the current business area, which is oral implantology. This is not favorable to radical innovation or business development. Therefore, lead users are good intelligence sources for the "keeping abreast" area rather than for the "looking beyond" area.

Comparison of TI elements at Straumann with the literature

Several elements presented in Figure 4.5 can be found at Straumann. Some of them are more or less established, for example ITI activities. Other elements are rudimentary at the present. However, none of these TI elements at Straumann are formalized, nor consolidated within a Technology Intelligence system. (See Figure 6.9.)

The TI process

This section is divided into two parts. The first part describes the designed screening process at Straumann, because this is a quasi-formalized process that is tangent to Technology Intelligence. In order to be thorough, the second part discusses the TI process along the lines of the TI activities presented in Chapter 4. There might be some redundancies, for example for information sources.

The screening process

The screening process (Figure 6.10) has been mentioned several times before. This chapter discusses inputs and outputs of the process and the process itself, as it is planned at Straumann.

Input consists of internal information/ideas (the process model does not distinguish between ideas and information; this is a weak point because information is neutral and ideas are normally in the context of the company, for example an innovation idea. Analysis criteria for them should therefore differ) from production, research, and employee feedback, and external information/ideas from various technology sources (universities, patents, competitors, literature, seminars, and fairs), customer ideas, and market research. No observation criteria are as yet defined, i.e. ideas and information enter the funnel more haphazardly rather than deliberately. This is reactive information-gathering. Consequently, there is a danger that nonrelevant information will be assessed, and potentially relevant information will not enter the screening process.

The process output is firstly a development project proposal, followed by a process where information about rejected ideas or project

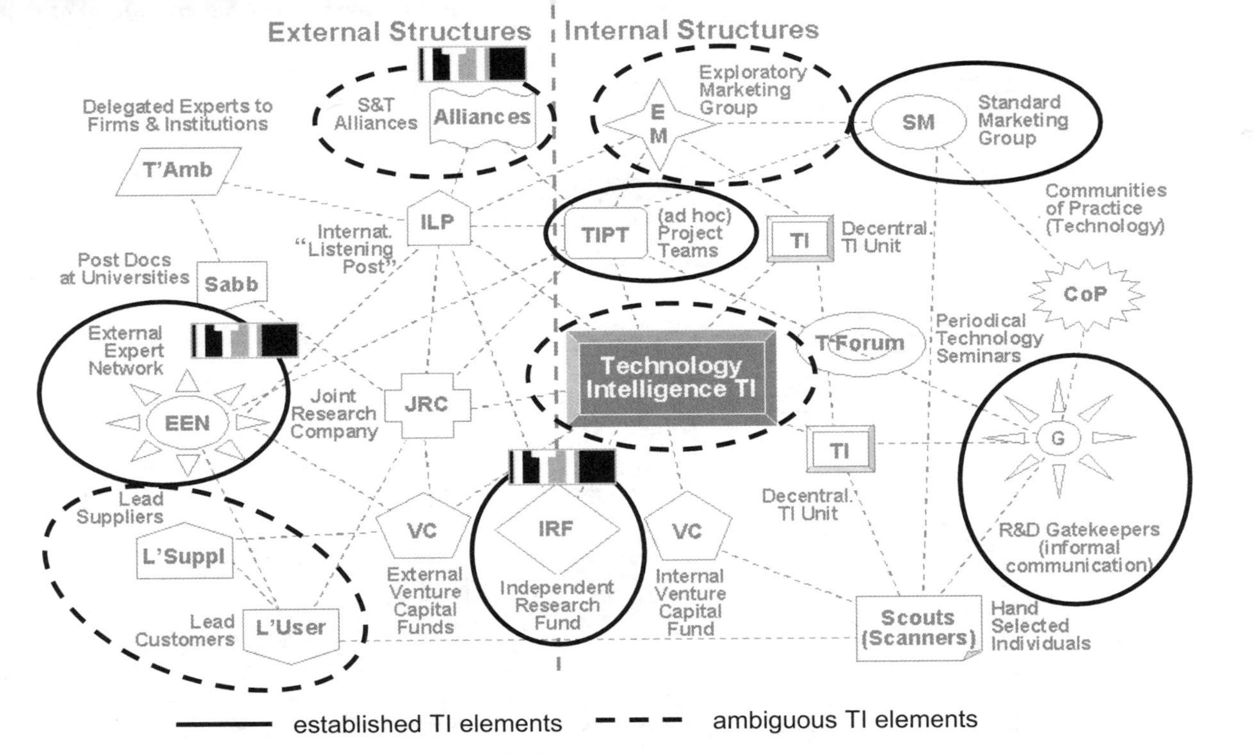

Figure 6.9 TI structures at Straumann compared to the literature

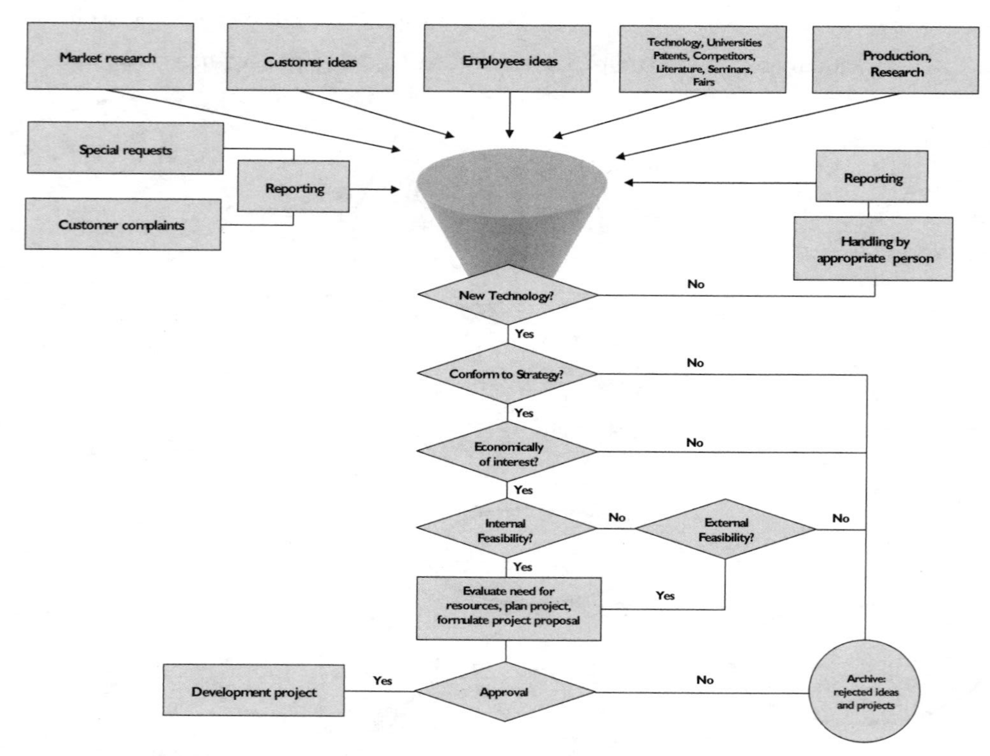

Figure 6.10 The screening process at Straumann

proposals is filed. Project proposals build a complete information base in order to enable a decision whether a project should be initiated or not. There is a clear yes or no decision. From this point of view the screening process is an intelligence providing process. The database with filed information about rejected ideas or project proposals underlines the learning aspect of this process. In fact, any analysis done during the process can be "recycled" at a later date whenever a similar idea or similar information needs to be analyzed. At the same time technologies can be tracked. An example would be that literature review signals a new material that might be interesting for oral implantology. A year later a university asks for cooperation on a clinical research study to test this material. If, for example, Straumann decides not to cooperate, but the university undertakes the study anyway, this might be a signal that a competitor is supporting the study. A few years later, if patent review shows a patent application, the new material could soon be introduced into the market. This example shows that the screening process delivers information not just for product development, but for strategic decisions too. However, again, the process is reactive to information entering the funnel. It would be dangerous to consider screening output results as adequate for strategic decisions.

Considering the process itself, an idea/information has to pass several criteria. These criteria are: Is this a new technology? Is it in line with strategy? Is it of economic interest? Is it internally or externally feasible? These criteria have not yet been detailed. In fact, these criteria hide some problems. Firstly, there is no difference between ideas and information. Secondly, the criteria are not applicable for just any idea or information, the focus is on technology. Thirdly, this is a sequential "stage-gate" procedure that could make an idea obsolete at the first gate. However, the same idea could be very interesting from the viewpoint of other criteria. Fourthly, the fit to strategy – very important and indispensable – is difficult to assess, because the strategy is not clearly formulated. Fifthly, in this early stage it is very difficult to estimate the economic (market) impact. Lastly, no communication logic (information pull/information push) has been defined. Thus, this process gives preliminary ideas how about the screening process could function. However, the process is not yet functioning maturely.

TI activities at Straumann

The content of this part is a discussion of TI activities at Straumann by means of the discussion framework introduced before.

Formulation of information need

How is the impulse for TI activities given at Straumann, and how is the observation area defined? Some needs are formulated more or less explicitly, for example the need for a decision support basis for product development, or explorative research studies. Some TI activities are also pursued proactively, responding to an implicit formulation of needs, for example during the daily project work. The observation area is almost exclusively limited to the field of oral implantology. Therefore TI activities focus on an "inside-out" perspective. Figure 6.11 shows TI activities at Straumann. A permanent change of the emphasis could not be observed.

Information collection

Who collects information and what are the information sources? The role of involved persons has already been discussed in the chapter

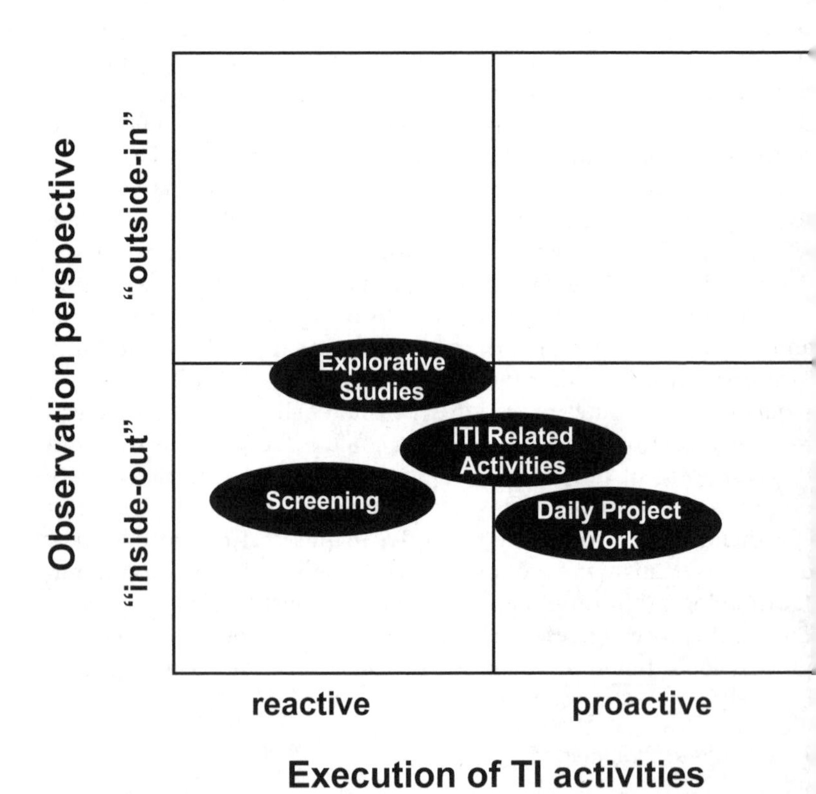

Figure 6.11 Observation perspective and impulse for TI activities at Straumann

describing the TI structures at Straumann. Some information sources which apply to information collection were listed in the screening process part and will also not be repeated. Comparing these and supplementary information sources to those presented in Figure 4.10, one can say that Straumann profits to some extent, from most of the sources that literature specifies. This implies that access is not categorically restricted to large companies. Whereas interorganizational cooperation, for example by joint subscription to a on-line database, could not be observed (with the exception of the ITI, which is almost an intraorganizational cooperation).

Information analysis

How is information filtered, integrated, and assessed at Straumann? To filter means to reduce the quantity and to assess the quality of information. This task is very well accomplished within the screening process. However, outside the screening process a great deal of information reaches numerous employees, sometimes several-fold. For this, there is no quantity reducing instrument, and no quality check happens in a coordinated way. It is up to each employee to filter the information she/he gets. The integration of information, i.e. to put the information into the Straumann context, is also guaranteed within the screening context and during daily project work. In turn this is more delicate with ITI activities, because the ITI context and the Straumann context are not necessarily congruent. The comments about filtering and integrating are also applicable for information assessment.

Straumann does not utilize technical analysis tools like trend extrapolation or the like, nor does the company revert to external experts to provide technical analysis. "Informal" tools such as intersubjective opinion forming are used sometimes, but there is no "analysis tool library" and there are no guidelines on how to use which kind of tool.

Information dissemination

How is the information flow logic organized at Straumann? In general, it is up to the interested person to get the information (information pull). Therefore, it would be important to have a maximal transparency on where to find the information. In fact, this is not yet resolved at Straumann: there is no appropriate "who-is-who" or file database. In turn, there are very short communication distances – the Technology Center is at a single site – and there is an open and informal communication climate, which is favorable to an efficient information flow. The emphasis is on oral communication. Lengthy written reports are avoided whenever possible.

Within the screening process, communication happens mainly by participation. However, the formal screening output consists of short written reports that can be tracked over a long period of time, as well as detailed (standard) project proposals.

Information application

How is the generated intelligence used? Since Straumann has no formalized intelligence system, it is difficult to make a statement about how "intelligence efforts" pay off. The fact that the early stage of innovations should be supported by the screening process lets us assume that the current situation was not satisfactory. There seems to be the desire for more decision support. From an organizational-learning view, the multidisciplinary and participative screening process is headed in a promising direction.

TI tools (methods and infrastructure)

Quantitative TI methods are not used at all at Straumann, and some qualitative methods are rarely in use. Compared to the numerous methods used in large companies (Table 4.3) the number of methods used remains low. Straumann undertakes benchmark studies, organizes expert panels, and leads some flexible expert interviews. There is in general a wish for a more systematic and coordinated use of both qualitative and quantitative TI methods. The technical infrastructure is favorable for TI activities. On-line access to numerous information sources is possible with access to the internet from every workstation. There is also a well-documented internal library. For communication and database purposes a modern platform (Lotus Notes) is installed. Also communication is stimulated by short distances and coffee corners.

Insights from the case study and action requirements

The previous chapter has shown that Straumann does not run a formalized Technology Intelligence system. Nevertheless some people or groups pursue TI activities more or less informally. This indicates the real need for TI. A first initiative to treat these activities more systematically is through the screening process. This process is planned but not yet implemented. The focus of the screening process is on product development, and thus on existing business; and it is short-term oriented. The observation of long-term trends, as postulated by the

Technology Center mission, is not guaranteed. However, the screening process strengthens the assumption that top management is committed to becoming better informed about information from the company's environment and committed to more formal processes.

Some existing structures (for example the ITI) and factors (for example, an open communication culture) are very favorable to build a TI system upon. Other structures and factors should be strengthened. For example, there is no "integrated technology management culture," and the business mission and strategy do not allow an observation focus or framework for intelligence activities to be deduced clearly.

To highlight insights from the case study, several positive and negative preconditions for an appropriate TI system are listed in Table 6.2.

Based on the theoretical background gained in Chapter 4 and the above-mentioned preconditions at Straumann, the following general **action requirements** for an appropriate Technology Intelligence system at Straumann can be formulated:

Table 6.2 Positive and negative preconditions for a TI system at Straumann

Positive preconditions for a TI system (strengths)	*Negative preconditions for a TI system (weaknesses)*
Clear commitment to increase the innovation power: this requires knowledge about future technological trends.	The business mission and strategy do not allow an observation focus for TI activities to be deduced clearly.
Top management supports the idea of a TI system.	The TI activities pursued are not institutionalized and coordinated.
An open company and communication culture is favorable for a TI system.	There is no shared understanding of an integrated Technology Management.
Existing structures are favorable, for example the ITI and its network of external experts.	The planned screening process has some weaknesses: exact content and persons involved are not yet defined, the process is very sequential and reactive, and the focus is restricted on the near future and thus, ... there is no basis that allows for capturing trends in the middle- and long-term future.
Preliminary initiatives for related processes have al ready been undertaken, for example the screening process.	
The technical infrastructure for information storage and communication already exists.	
Numerous information sources are at Straumann's disposal, for example an expert network, an extensive library, internet access for everyone, etc.	

- **Definition of a Technology Intelligence framework:** In consideration of existing TI activities and the screening process, an overall Technology Intelligence framework that determines an observation scope should be defined.
- **Definition of an integrated technology management context:** Following the insights from theory, it seems to be favorable to design a participative, decentrally conducted and coordinated TI system. For this, a superordinated technology management context should be defined.
- **Reorganization of the screening process:** In order to make the screening process work as a part of the Technology Intelligence system, the process should be defined in detail and roles should be assigned.
- **Completing the Technology Intelligence system:** Next to the screening process, other elements of an appropriate Technology Intelligence system should be defined in order to cope with the defined framework.

One **restriction** limits solutions to these action requirements: **no additional personnel** should be employed for the Technology Intelligence system, i.e. any activities should be based on existing employees.

Technology intelligence solutions for Straumann: elements

This chapter presents different elements of a solution for a Technology Intelligence system at Straumann. It has to be remembered that this solution reflects business reality, and therefore is a trade-off between the academic state-of-the-art and pragmatism. Action research allows such a procedure.

The elements of this solution were jointly elaborated during a time period of almost two years. Therefore, this action research includes information and events in the company until autumn 2001. Changes after autumn 2001 are not considered.

The author of this book spent in total approximately 80 full days in the company for:

- *Interviews*: Interviews conducted during this action research case varied broadly. The interviews can be grouped as: Structured interviews which were conducted to build a broad understanding based on detailed questions and to triangulate data. "Open" interviews focused on one defined topic, but were without structured questions. Open discussions which led the author to a deeper understanding of the company context and provided the author the opportunity to compare theory with people's behavior. This third type of interview allowed Straumann employees to tell "stories," which was important to learn about their experiences. Duration of interviews varied, by the interview's nature, from a few minutes to some hours. Interviewed persons were from all functions of interest to Technology Intelligence, i.e. top management, the Technology Center and production.

- *Workshops*: Six workshops built the basis for taking action. Workshop topics were about technology management, Screening and the Opportunity Landscape (cf. later in this chapter). Workshop participants ranged from top management to R&D staff. This participative process jointly generated knowledge about a specific topic. By these means, an optimal solution for a problem could be found based on intersubjectivity. The workshops were mainly coached by the author of this book, sometimes assisted by an external professional coach (this coach has worked with Straumann for a long time). Workshop duration was either a half or a full day.
- *Presentations*: Presentations helped the author to diffuse the generated knowledge to a specific audience, which was the board of directors, the executive board, middle management, or staff. The difference between presentations and workshops is that presentations seldom generate new knowledge.
- *Meetings*: In the sense of action research, i.e. observation of the influence of various actions, the author assisted and even led some "operative" meetings, especially the monthly screening meeting (cf. later in this chapter).
- *Desk research*: Mainly corporate annual and monthly reports, internal memos and protocols, company journals, and project reports.
- *Analysis work*: Analysis work consisted of working on the collected data, i.e. structuring and analyzing the data, and developing concepts.

Thus, this chapter explains, in addition to the elements of the solution of Technology Intelligence, what led to the formulation of these elements. According to Yin (1988: 139), the solution is presented in a "suspense" structure, i.e. the outcome of the action research study is presented first, followed by the development of the explanation of this outcome, which are elements of TI.

In order to be comparative with the situation at Straumann before this action research, and with the existing literature, this chapter closes with a short discussion of the TI system by means of the adapted value-chain model.

Elements in a holistic context

The jointly elaborated solution for a TI system consists of six main elements. These are business mission and strategy, the technology management group, an external expert network, the remodeled screening process, the Opportunity Landscape, and a common IT platform (Figure 6.12). It has to be remembered that this TI system runs in parallel to and interacts with general business processes, which means with other "management systems" (for example R&D, HRM, patents, quality management). Thus, as other management systems do, the TI system supports the innovation and technology potential, which is symbolized by the dental implant in Figure 6.12. In fact, the focus of innovations at Straumann is exclusively on products. This view neglects other forms of technology (= R&D outcome) use, for example selling and licensing out technologies (Escher 2001: 4). Such alternative forms are commonly understood as technology marketing or IP business. However, since the business strategy of Straumann does not allow for activity in this field, the IP business view is not integrated into this book.

In fact, the two core elements of the TI system are the **screening process** and the **Opportunity Landscape**. They are complementary: the goal of the remodeled screening process is still to lead to effective and efficient decision-making before product development. Thus, the focus of this process is mainly short term and on the actual business ("keeping abreast"). The Opportunity Landscape, in turn, covers the "looking beyond" area, which means that activities within the Opportunity Landscape are mid- and long-term oriented and cover, next to the actual business, any field of interest which is relevant to Straumann. Today's activities within the Opportunity Landscape may become tomorrow's activities in the screening process.

The business mission and strategy are defining elements of the Technology Intelligence system: the TI mission and goals are directly derived from it. However, since insight from TI activities may influence the business mission and strategy, the latter is also considered part of the system.

The **technology management group**, the **external expert network** (ITI), and the common **IT platform** are supporting elements. The first

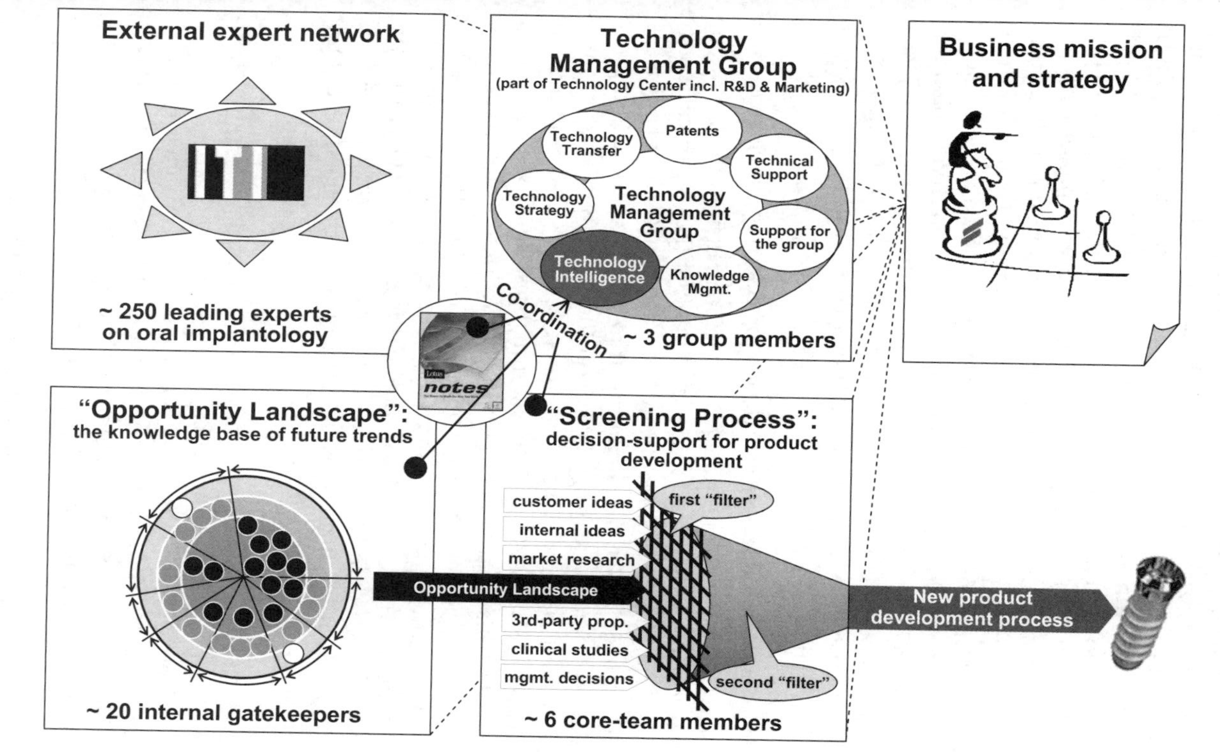

Figure 6.12 Technology Intelligence elements at Straumann in a holistic context

is the coordinating element of the system. There are no specific TI activities done directly by the technology management group (in the sense of a centralized Technology Intelligence unit). One of the group's tasks is to manage (design, direct, and develop) the screening process and the Opportunity Landscape. The ITI in its function as external expert network is a primary source of information. In addition to this, some TI activities, for example explorative technology projects, happen entirely within the ITI, but are at Straumann's disposal. Thus, this external expert network is considered an important element of the TI system at Straumann. The common IT platform (based on Lotus Notes) is a very important communication and information storage platform, and therefore is also considered an element of the TI system.

Details about these six elements of the TI system at Straumann are the content of the next chapters. The discussion order is chosen in order to be most comprehensive.

Business mission and strategy in relation to Technology Intelligence

The business mission and strategy of Straumann defines the business in which the company wants to act. In addition to this, a particular technology field, in which the company wants to strengthen its knowledge, is highlighted. In fact, the commitment to becoming active in this particular technology field has been considered in the latest business mission and strategy paper only; there was no such commitment before, i.e. in 1999/2000 when the initial case study was conducted. Essentially, this commitment seems to be provoked by uncertainties in the screening process and the Opportunity Landscape, as we will see later. Furthermore, Straumann strategy states a clear commitment to technological leadership. This statement is strengthened by the R&D mission, which claims to keep and increase the innovation power.

The implication of this business mission and strategy in relation to Technology Intelligence is twofold:

- The TI system should allow proactive TI activities observing short-, medium, and long-term trends.
- The observation focus of TI activities is not restricted to today's business. However, with the exception of the specified new technology field, no clear statement about the business/technology field observation areas has been made.

The observation area is represented in Figure 6.13.

From this new definition, a TI mission can be deduced. Since the existing screening goal already covers a part of the intended observa-

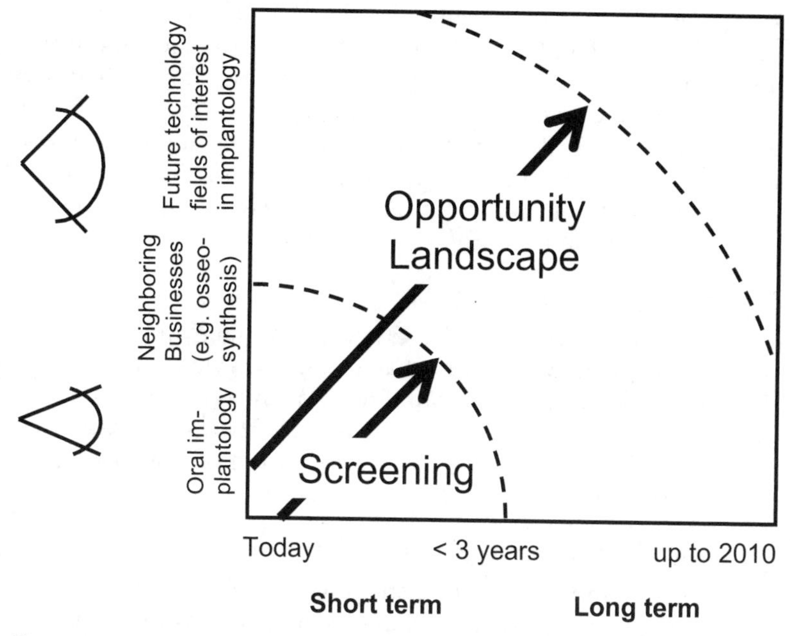

Figure 6.13 Observation area of TI activities at Straumann

tion focus, it has been adopted for short-term trends in the existing business. Thus, the first **TI mission and goal** is:

> *to lead to effective and efficient decision-making before product development.*

The second mission and goal completes the rest of the intended observation focus. This means medium and long-term observation in neighboring and new businesses/technology fields, and areas that are not covered by the screening process. This part will be covered with the Opportunity Landscape. In fact, following the R&D mission to increase the innovation power, the second TI mission and goal is:

> *to lead to effective decision-making for future innovations by building a knowledge base of present and future facts and trends.*

In contrast to the influence of business mission and strategy on other elements, insight gained from TI activities, especially within Opportunity Landscape activities, may in turn, influence business mission and strategy. Whenever strategic decisions are adapted, knowledge from the screening process and the Opportunity Landscape may build a decision support. For example, Straumann decided to explicitly mention a specific technology field in their new business mission and strategy. This field caused several problems previously, because it was not clear whether observations should be done proactively in this field or not. However, this relationship is difficult to prove. But this example shows the interactivity between business mission and strategy with other TI elements.

The technology management group

The situation described in the case study "Straumann 1999" has shown two structural problems that are unfavorable for a TI system: there is no common understanding of an integrated technology management, and there is no optimal structural unit that could coordinate TI activities. This led us to initiate a technology management group.

Justification for the technology management group

In fact, considering the existing structure of the Technology Center (Figure 6.5), there are as many staff positions – and with them very specific tasks – as functional units. This seems to be exaggerated and implies some dangers:

- One staff function means one single person: A function (for example patents) can reach dimensions that one single person cannot manage (for example litigation needs skills other than simple patent administration).
- In turn, one single person often means one single task: Other important technology management tasks (for example, technology strategizing) are likely to be missed within such an organization.
- Isolated bunker: Staff employees are often afflicted with missing acceptance and support from other units. This often excludes them from daily business and "real business problems." In addition, communication among staff employees, and between staff employees and other employees, is often complicated because they are only partially occupied in the same projects.
- Overtaxing with special tasks: Some special tasks (for example, set-up of an IT database) are often allocated to these staff functions. This may inhibit the normal tasks that are part of their day-to-day responsibilities.

The existing literature approaches staff functions quite skeptically. Global organizations in particular try to redistribute headquarter planning staff to business units. However, advantages and disadvantages of staff functions at the business unit level are not given. There is little insight into how SMEs handle staff functions. It was felt that bundling technology management tasks into one technology management group made sense (Figure 6.14). An exception was clinical research. This function is still considered an individual staff function because it is a "product development process" that is parallel to research and development.

Attention has to be given to the fact that this technology management group should not operate all technology management tasks individually. This would be contrary to the understanding of an integrated technology management. On the contrary, the technology management group should be seen as a service group to other functional units and management. Thus, the technology management group fulfills various coordination tasks.

Tasks in the technology management group

Independent from the personnel question, some tasks were more sensibly defined. Some tasks were adopted from the former staff functions, other tasks (Technology Intelligence, technology strategy, and knowledge management) complete former tasks (Figure 6.15). This is not a definitive or complete status but a first attempt and trade-off to unify Straumann's real needs and range of action with the most important technology management tasks. Therefore, the task definition of the technology management group remains a dynamic process, for example "Patents" could be enlarged to "IP business," including purchasing/selling and licensing of technologies.

Figure 6.15 lists some key words that are possible contents of the tasks. Again, this is not a definitive list and corresponds to a combination of the state-of-the-art, currently pursued actions and a "should be" situation.

The task of particular interest is Technology Intelligence. It has been mentioned before that there are, *a priori*, no TI activities directly affected by the technology management group. This group has a managing and especially a coordination function with regard to the various TI activities. Therefore, it is the group's responsibility to manage the screening process and the Opportunity Landscape.

Structural organization and philosophy of the technology management group

In the technology management group one task is not strictly assigned to one single person. The members of the group fulfill the different

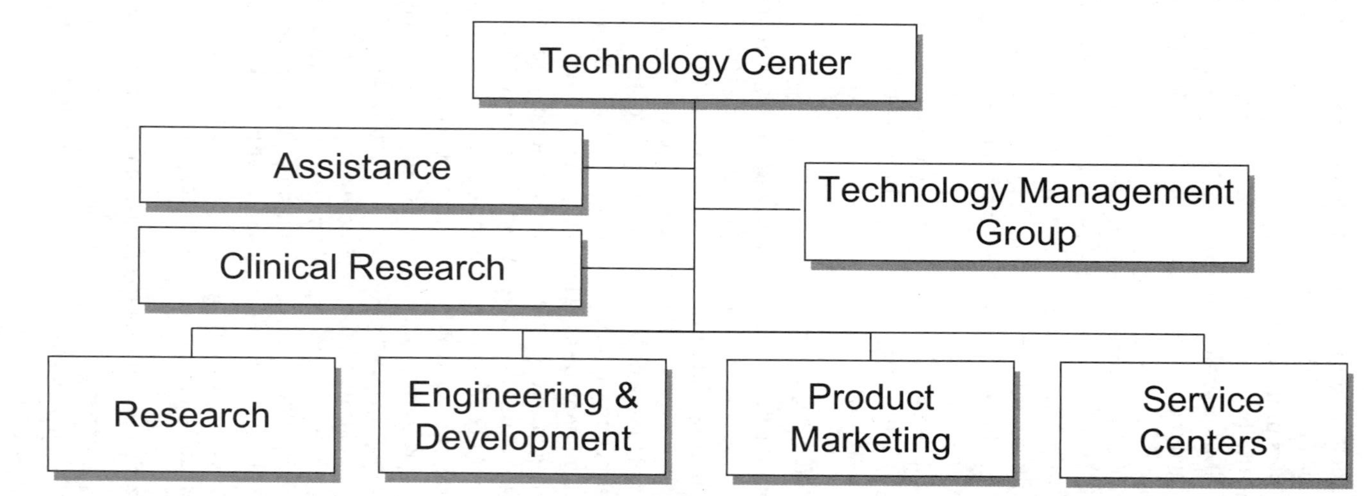

Figure 6.14 New organization of the Technology Center including the technology management group

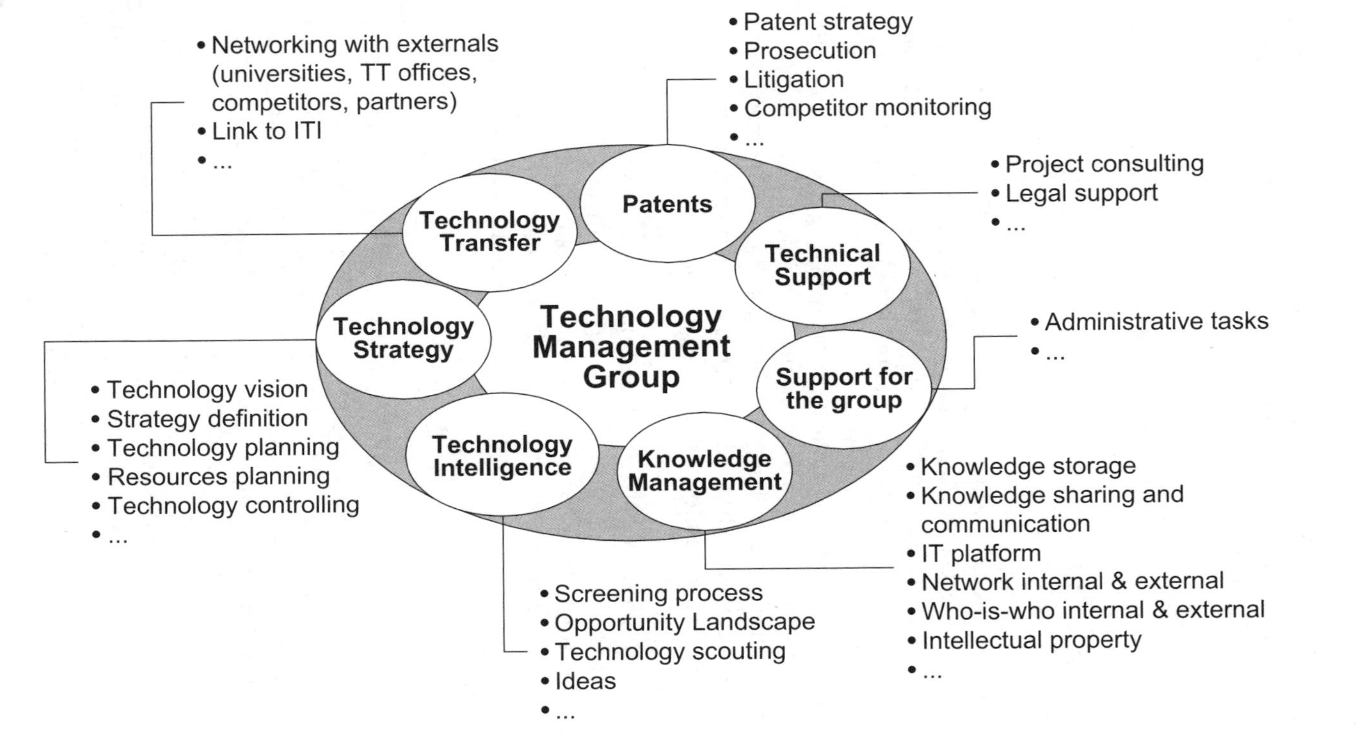

Figure 6.15 Tasks and content of the technology management group

tasks as a consortium. Of course some people are specialized in some tasks, but whenever possible a task should be shared by at least two people. Benefits of this kind of organization are:

- more tasks than people can be considered,
- knowledge is shared in order to avoid the problem of "leaving experts,"
- different viewpoints are considered,
- bottlenecks are reduced.

Nevertheless, the responsibility should be clear. Such an organizational form is very flexible and can change the main focus rapidly. This is very important because in the future Straumann will be more and more confronted with very complex problems from inside and outside the company.

From a present viewpoint, the technology management group could consist of the people depicted in Figure 6.16. The ticks in Figure 6.16 are provisional. Nevertheless, they show that the existing staff are not sufficient to complete the tasks. Therefore a "freelancer concept" seems to be a promising way to close these gaps. Also other tasks might be considered gradually.

How is the group organized? Management of this technology management group is cooperative, which means that there is no hierarchy in it. Every single group member is responsible for designing, directing, and developing the activities of the group. However, there is a "*primus inter pares*" (first among equals) for administrative tasks like budget planning.

External expert network (ITI)

The ITI and its relation to Straumann have been mentioned several times before. Also the very important role of the ITI with regard to Technology Intelligence has been highlighted in the case study "Straumann 1999." Thus, the ITI is considered an important element of the TI system at Straumann. This element functions as a primary information source, as well as functioning for some analysis activities. Figure 6.17 accentuates how important the formal and informal networks are, especially the extended networks to universities.

Contact with this external expert network is particularly interesting for the two main TI elements at Straumann, which are the screening process and the Opportunity Landscape. As we will see later, members of the ITI may be involved in further investigation of the screening process, and they may be very valuable contributors to

	Technology Strategy	Technology Transfer	Technology Intelligence	Patents	Knowledge Management	Technical Support	Support to the group	
Person A	(✔)	✔		✔		✔		
Person B				✔		✔		
Person C			✔		(✔)			
Person D							✔	
"Freelancer"	(✔)		(✔)		✔			

Figure 6.16 People in and organization of the technology management group

the Opportunity Landscape. Thus, some ITI members will have access to the common IT platform.

The new screening process

The initial situation at Straumann with regard to short-term oriented Technology Intelligence activities was already promising. The design of the screening process had been initiated, thus there was a positive precondition for new concepts. The "new" screening process is an evolution of the "old" one, because the latter had obvious weaknesses. The "old" screening process was still in the design phase, a situation which was favorable to changing the process without strong opposition.

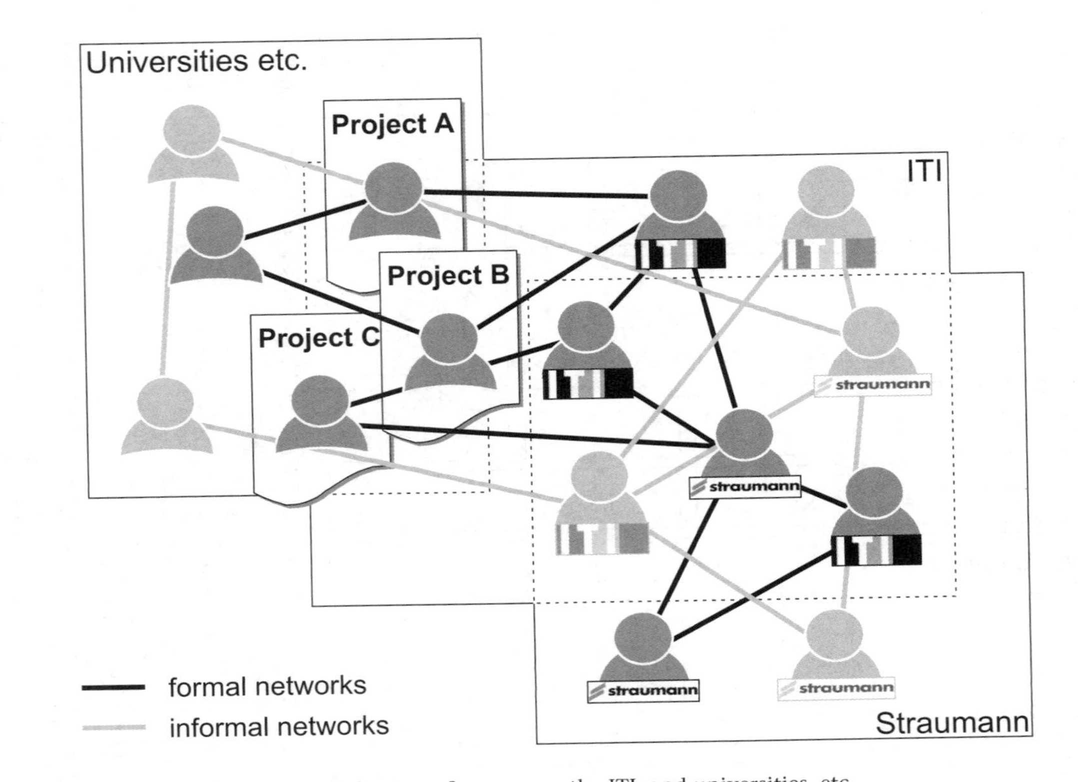

Figure 6.17 Formal and informal networks between Straumann, the ITI, and universities, etc.

The structural organization that conducts the screening process will be presented first. The old process already envisioned one person – the screening manager – who was to be employed specifically for this task. The new screening process will then also be explained, followed by practical experience gained during a period of about one year.

Structural organization of the screening task

The concept of a multidisciplinary approach has been adopted from the initial situation. In addition to competencies from various functions, the direct link to top management is assured through the head of the Technology Center as another screening team member.

Hence, the screening core team includes in total seven members (Figure 6.18):

- the screening manager,
- the head of the Technology Center,
- the head of research,
- the head of development,
- the head of clinical research,
- a representative from product marketing, and
- a representative from the market unit.

Most functions are represented by their department heads, which makes this screening core team a competent and decision-authorized committee. The presence of product marketing and market representatives mitigates what might be an excessively technological view.

The screening core team meets twice a month. The first meeting is the formal screening meeting where decisions are made. The meeting duration is about two hours, and participation of the screening core team members is mandatory. The screening manager coaches this formal meeting. In between these formal meetings some core team members meet for the informal screening lunch. Participation is voluntary. The goal of this lunch is to discuss "screening inputs," and if possible to ante-date decisions. This screening lunch was initiated because the time span of one month is too long for treatment of some inputs.

If a screening input needs further investigation (cf. the screening process), a virtual *ad hoc* team can be put together. This is a time-limited team consisting of some screening core team members and/or other internal experts and/or external experts. The composition of the team depends on the competencies needed. External experts come from Straumann's formal and informal networks, especially from the

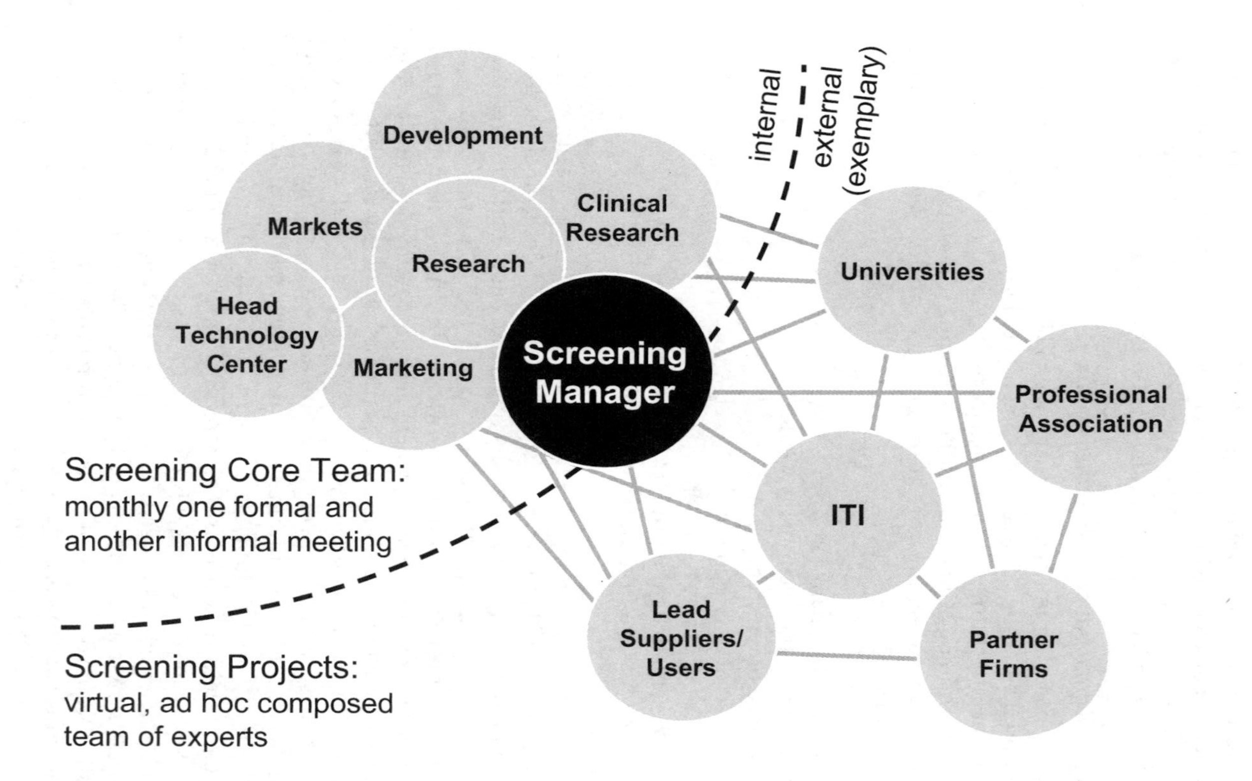

Figure 6.18 Structural organization of the screening process

ITI. Again, the screening manager coaches this virtual *ad hoc* screening project team.

Special attention has to be given to the screening manager. As mentioned before, he is the coach of the screening core team and *ad hoc* screening teams. This makes him the common denominator of any screening activities. Thus, he is responsible for the effective and efficient execution of the screening process. For example, since he manages the screening process, he is also responsible for the development of the process. For this, the screening manager ideally has managerial skills in addition to his technical knowledge within the area.

The new process design

The screening process design has been remodeled in order to strengthen the weaknesses of the planned "old" process. the strengths of the old process have been maintained whenever possible.

The new screening process is separated into two stages which differ in depth of examination. There is no step-by-step evaluation which covers the unavoidable danger of rejection of a promising idea because of one single criterion, but each of both steps consists of a holistic evaluation of the screening input. The two steps correspond to two filter functions of the process: selection and investigation (Figure 6.19).

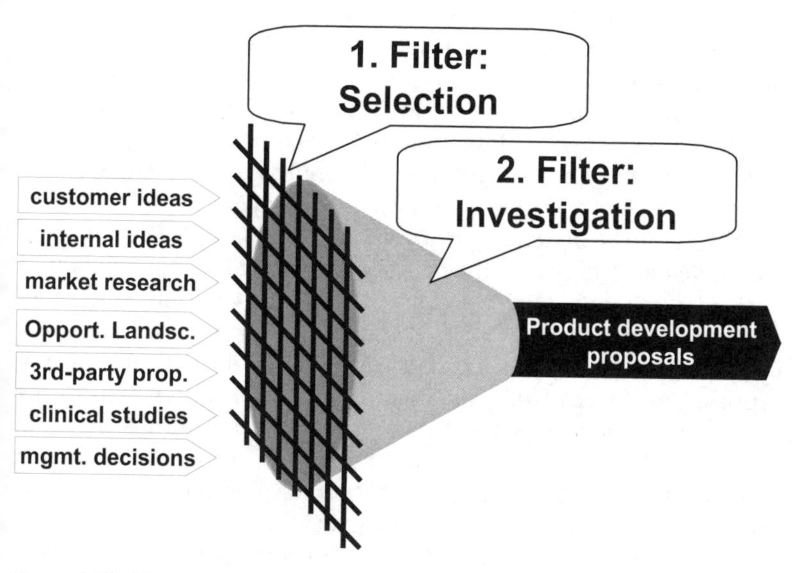

Figure 6.19 The two-stage screening process at Straumann

The screening input is upgraded with the input from the Opportunity Landscape, which simultaneously covers the proactive part of the screening process. All other forms of input conceived in the old process are still valid: customer/internal ideas, market research, third-party proposals, clinical studies, management decisions, etc. The screening input still consists of ideas or information. However, as we will see later, the process has been designed in a manner that allows both types of input to be handled.

At first, any screening input is directed to the screening manager, who in turn redirects the input to the most competent screening core team member for a preliminary, spontaneous assessment. Then the screening input is presented (neutrally) and discussed by all screening core team members during the monthly screening meeting. After this short discussion among the multidisciplinary team, a collective decision is made whether the input is of interest to Straumann or not. This is the first filter, which is **selection** of various screening inputs. There is no explicit criterion for this selection. However, the multidisciplinary team guarantees a best-funded decision because they have their implicit technical, economic, strategic, and organizational criteria which are based on the business mission and strategy, and on their daily tasks as head of a business function. The advantage of the absence of explicit criteria is a fast and straightforward processing of any screening input.

If the screening input is considered interesting to Straumann, it becomes a screening project and enters the further **investigation** stage, which may last from one day to several weeks. The screening input is subject to a detailed analysis from four perspectives: technical, economic, strategic, and organizational. There is a detailed criteria catalogue for each aspect. For example, in this stage a patent situation analysis has to be performed. However, this catalogue should always be considered with respect to the nature of the screening input, because a particular criterion might not make sense for two different cases. Detailed analysis needs specialized knowledge. Therefore the screening core team designs a virtual *ad hoc* expert team (cf. the structural organization). The result of this investigation, in turn, is presented to the screening core team a second time during another formal meeting. Based on these results, the multidisciplinary team decides again whether the screening project should become a product development project or not. This represents the second filter.

The output of a positive decision is a **project proposal** which has to be approved by the executive board. This finishes the screening process for this screening input. If the project is accepted, any information and documents are transferred to the new project leader.

Whenever a screening input is refused after one or both filters, all information and documents are stored in a database in order to be able to reconstruct the "history" of the screening input. Therefore, the output of the screening process, next to project proposals, is analyzed information, which is intelligence. The database is a well-structured library of useful information stored for future use.

The manner in which a screening input passes through the screening process is depicted in Figure 6.20. This process is a standardized part of Straumann's quality management (QM) process. Therefore, there are standardized forms with quality numbers. However, quality performance measurement has not yet been defined.

Implementation of the screening process

The screening process implementation was a part of the superordinated Technology Center project. Screening was considered an important function of the Technology Center. Autumn 2000 was the inauguration of the Technology Center; thus, the screening process was supposed to be implemented at the same time. However, within the screening subproject, no clear project goals and no exact milestones were set initially. The state of the screening implementation project seemed to remain at the state as it is described before. In retrospect, the major problem was the lack of project champions. Thus, the head of the Technology Center asked an external expert (the author of this book) to consult. This was a kind of relaunch of the project. The result corresponds to the new process design and the structural organization as it is described in the last two chapters.

The implementation (relaunched) project had two champions: first, the external expert was responsible for designing the screening process, as well as for project management of the implementation. The latter did not cause problems of acceptance because of the expert's external status. In fact, the external status was also favorable in some situations because he was independent of "internal politics." Also, the external expert got strong support from the head of Technology Center, who was the "political champion" or "sponsor" of this project. Other project team members were the heads of research, clinical research, development, and product marketing, which correspond more or less to the screening core team members.

Since all involved person knew the "old" process design, a half-day workshop was initiated at the beginning to create a common

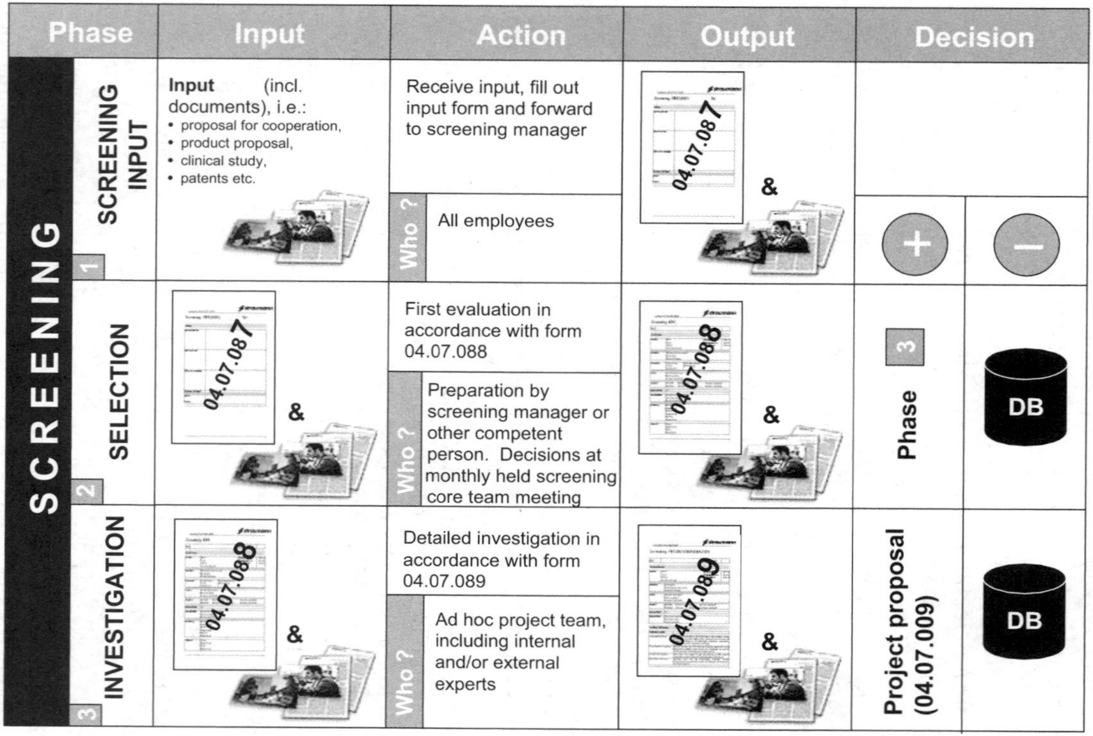

	Phase	Input	Action	Who ?	Output	Decision	
SCREENING	1 SCREENING INPUT	**Input** (incl. documents), i.e.: • proposal for cooperation, • product proposal, • clinical study, • patents etc.	Receive input, fill out input form and forward to screening manager	All employees	04.07.087 &	+	–
	2 SELECTION	04.07.087 &	First evaluation in accordance with form 04.07.088	Preparation by screening manager or other competent person. Decisions at monthly held screening core team meeting	04.07.088 &	Phase 3	DB
	3 INVESTIGATION	04.07.088 &	Detailed investigation in accordance with form 04.07.089	Ad hoc project team, including internal and/or external experts	04.07.089 &	Project proposal (04.07.009)	DB

Figure 6.20 The screening process as a standardized QM process

understanding of the "new" process design and to define the implementation scope. In this workshop, the project team decided to start directly with the new process design with concrete screening inputs, and then to make corrections "on-line." One adaptation, for example, was the introduction of the screening lunch, which is an informal meeting to speed up communication. In a first stage the implementation scope was restricted to Straumann in Switzerland in order to test the process. In a second stage, general managers of Straumann subsidiaries were informed. After some months, a new screening manager joined the team. Therefore the role of the expert champion shifted from the external expert to an internal Straumann employee.

The implementation of the screening process did not follow classic project management rules: it was rather fuzzy. However, the overall deadline – Autumn 2000 – could be met. Before implementation, ideas of how the process should be designed already existed. But since details were defined and adapted during implementation, one cannot strictly separate designing and implementation. The designing and implementation processes were highly participative, thus special training of "screening workers" was not necessary.

Practical experiences with the new screening process at Straumann

Straumann is in a favorable situation: a lot of ideas that correspond to current trends come automatically to the company through third-party proposals because Straumann is well known worldwide in the MedTech industry and regionally/nationally known outside the MedTech industry. In both cases, Straumann is respected for its innovative reputation and financial strengths.

This reactive input collecting is promising but not enough. Other trends should be "detected" through proactive investigations or observation. On the one hand, the Opportunity Landscape is supposed to give proactive input to the screening. On the other hand, further investigation during a screening project leads to new insight that could become a new screening input.

These screening projects have an auspicious side-effect: because most screening projects need an assessment of the competitors' situation, Straumann is forced to observe competitors regularly. This is also true for being up-to-date in literature and patents. Investigations give Straumann additional external expert contact that goes beyond the contacts to the ITI.

A numerical representation of the one-year screening process practical experience is shown in Figure 6.21.

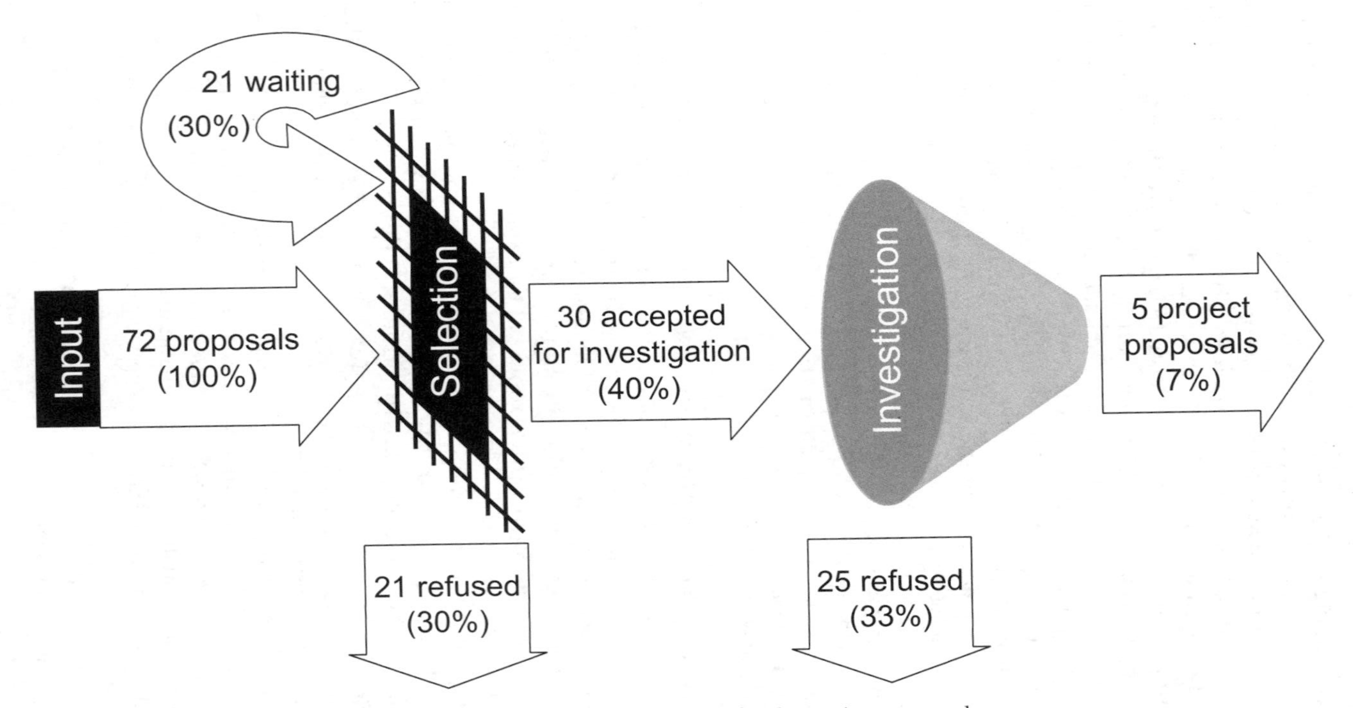

Figure 6.21 Practical experience with the screening process: input–refusals –project proposals

There were about 72 inputs, which seems to be few. A rough estimation of numbers of expected inputs was about 150. A major problem could have been that the screening process was not recognized throughout the company, and subsidiaries especially were not initiated adequately. Selection seems to work well for inputs on which decisions could be made. But there were still 21 proposals in a waiting position. In most of the cases no decision could be made because the situation regarding business mission and strategy was unclear, or because the team member in charge did not attend the meeting or was not well enough prepared. The former was attenuated with a revision of business mission and strategy. This problem was partly solved with the introduction of the informal screening lunch in-between the formal screening meetings. The 30 screening projects for further investigation turned out to be varied. Some of them took a day to evaluate, others took several days. Correspondingly, the financial resource use was also varied. Finally, 5 project proposals (= 7%) emerged from the initial 72 screening inputs. Whether this number is high or not is difficult to assess because there is no data from previous experience for comparison. A benchmark with other companies running a comparable process would probably be helpful.

In fact, the question about the amount of output provokes a performance measurement system. There is no such performance measurement for screening installed yet at Straumann. Literature does not provide adequate measurement methods. However, at Straumann some facts let us assume a satisfactory performance (with exception to the fact that apparently not enough input enters the process). Concerning measurement of effectiveness, for example, the subjective perception of top management seems to be favorable for the newly designed screening process because an equivalent budget has been maintained for the succeeding year. However, because of the very different nature of screening inputs, and because of many influencing factors other than the process design, a sound and significant performance measurement seems to be doubtful.

Another point of interest is human resources. When the idea of a screening process appeared for the first time, Straumann intended to occupy a screening manager full time, in addition to time spent by other involved persons. Today this person's activities demand about 50%. Screening core team members spend about 6 to 8 hours a month on screening, which is about 25% of the work time for one person. This vague estimation shows that the time spent on the screening process is within the "full-time limit." This account does not include the time gained by avoiding redundancies. Comparing the time resources used by this screening process with the wasted time when more than one person evaluates any input that is submitted to Straumann, one can say that in total there is probably an economy of human resources. For running costs, Straumann budgeted about US$40,000 in 2000, of which just US$12,000 effectively was spent. In addition to this, another US$12,000 was used for database set-up. Since hardware and software were already available, investment costs remained relatively low.

The participative and multidisciplinary screening process presents another benefit in the form of the internal network effect. During these formal and informal meetings any knowledge is transferred. This stimulates transparency for tasks other than screening – for example, project work. Also, first-hand information and experiences are rapidly diffused among employees, which is an organizational learning effect.

The Opportunity Landscape®

Because there were no systematic activities that allowed Straumann to accomplish the second TI mission ("to lead to effective decision-making for future innovations by building a knowledge base about present and future facts and trends"), and because no appropriate concept for SMEs to do so could be found, either in practice or in the literature, the author of this book designed a novel concept: the Opportunity Landscape®.

This section intends to describe firstly the concept in a neutral and rather theoretical way, and secondly, to show how the Opportunity Landscape works at Straumann. In order to be most comprehensive, a step-by-step discussion about design and implementation is presented in parallel. In fact, the discussion about design and implementation is very difficult to separate because the Opportunity Landscape is an interactive and participative concept. To explain the result (the design), the way to this result has to been explained (its implementation).

Goals of the Opportunity Landscape

According to the TI mission, the main goal of the Opportunity Landscape is to identify and anticipate relevant present and future facts, trends, and developments in the company's technological environment by constant and systematic observation. This knowledge is stored and maintained in a corporate knowledge base. The idea is to link these observations to business mission and strategy, and vice versa.

The concept of the Opportunity Landscape

The Opportunity Landscape is *a priori* designed for SMEs. Therefore, special attention has to be given to resources, i.e. to limit manpower needed. Thus, actions within the Opportunity Landscape are based on and shared among current employees. This is an "organizational intelligence" approach and may support, therefore, any business activities as a corporate knowledge base about future facts and trends. The concept of the Opportunity Landscape is the subject of this section.

Definition of relevant strategic fields and issues

In the beginning, a list of issues has to be agreed upon. These issues are supposed to reflect topics relevant to the company. The character of an issue is not defined *a priori*, so the Opportunity Landscape is not restricted to technological issues. However, for a technology-based company most of the issues will have a technological character.

There are two basic approaches to define the list of issues: top-down and bottom-up. The first approach takes the business mission and strategy as the starting points to define strategic observation fields, within which several relevant issues are allocated. This task should be accomplished by the Executive Board and the Board of Directors in order to reflect the best business mission and strategy. In the second, the bottom-up approach, people, including executives from different departments, for example R&D, Marketing, or Production, with correspondingly varying views, are brought together in strategic workshops. Creative methods like brain-writing, where each person is requested to define at least 20 important issues of interest, are used to come to an agreement. The results are then discussed, consolidated, and grouped into strategic fields that have to be checked for alignment with business mission and strategy. This is an iterative process. The use of external consultants may be appropriate for workshop moderation or for technical input.

The choice of the top-down or bottom-up approaches depends on the exact scope of the Opportunity Landscape and on the company's organization and culture.

Definition of observation depth

Not all of the issues found are of the same importance to the company at a given time. Therefore, the Opportunity Landscape holds three focus areas: players, substitutes, and juniors. Players' issues are constantly and intensively observed and reported in detail; such knowledge is needed "today" for imminent decisions. Substitutes' issues are attended to at regular intervals; information should be available in medium-terms. Juniors' issues are retained, but not regularly monitored.

This classification of observation depth gives the Opportunity Landscape an additional dimension.

Visualization of the Opportunity Landscape

It is important to visualize the Opportunity Landscape in order to give it a face and to obtain an overview of all the issues. This improves the transparency of the activities. In addition, schematics are excellent communication tools. In general, visualization requires three qualities: simplicity, persistence, and completeness. Simplicity measures the degree to which the visualization is immediately understandable, persistence measures the propensity for the visualization to linger in the mind of the beholder, and completeness measures the extent to which all relevant information in the data is depicted.

An example of the visualization of the Opportunity Landscape, which presents strategic fields, issues within the fields, and observation depth, is depicted in Figure 6.22.

Definition of "gatekeepers"

The most important resources of the Opportunity Landscape are people. This concept is based on a philosophy that is derived from Allen's (1986) **gatekeeper phenomenon**. For a better understanding, this phenomenon is briefly described.

During his study of many years in R&D departments of numerous firms Allen gained significant insight into communication behavior:

- Most information for innovation stems from external information sources.
- Successful scientists and engineers communicate more frequently than nonsuccessful ones with both people from inside and outside project teams.
- Some scientists and engineers are more intensively involved in a communication network than others. In addition, the former present above-average contacts to the company's environment. Therefore, they are referred to as gatekeepers for outside information (Figure 6.23).
- The gatekeepers can be easily recognized on the basis of three characteristics: (a) the gatekeeper is a highly technical performer, (b) about 50% of gatekeepers are first-line supervisors, and (c) technical management can guess accurately who gatekeepers are.

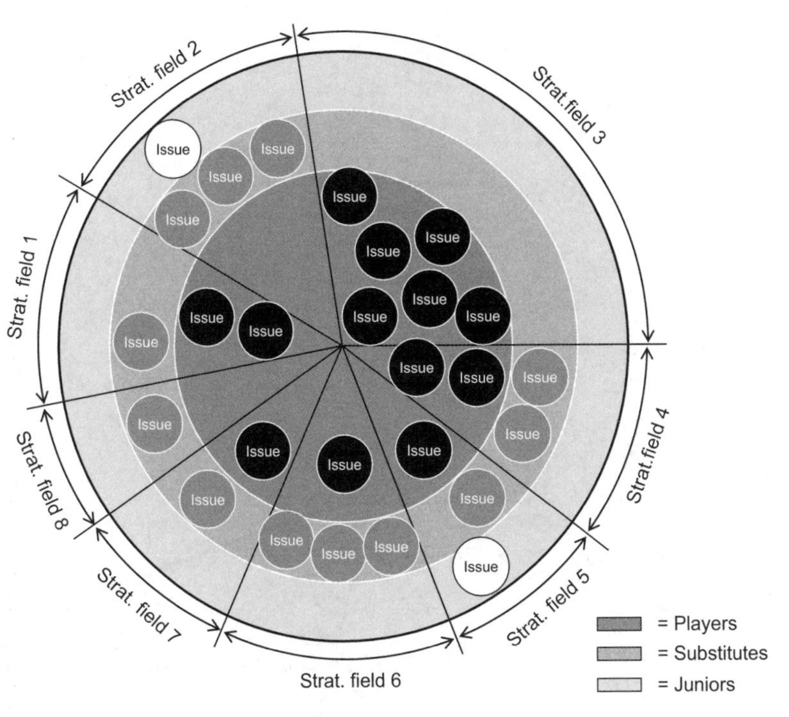

Figure 6.22 Visualization of the Opportunity Landscape at Straumann; illustrative definition of "gatekeepers"

- Already a short distance separating gatekeepers considerably decreases communication flow.

Back to the Opportunity Landscape: for each defined issue, the most competent person has to be identified and is then referred to as the gatekeeper. Normally an expert who has been known for some time to be familiar with all aspects of an issue is labeled as the gatekeeper for that issue. Sometimes, though, hidden experts have to be discovered first. In fact, the gatekeepers in the Opportunity Landscape are "instilled" gatekeepers, and thus differ slightly from Allen's definition. They actually do not necessarily present all the gatekeeper's skills described before, but they are supposed to. So the tables are turned if a defined gatekeeper does not present the necessary characteristics – she/he has to be trained. However, the goal is to match the best instilled gatekeepers with "real" gatekeepers. The choice should be done by line managers and the Executive Board.

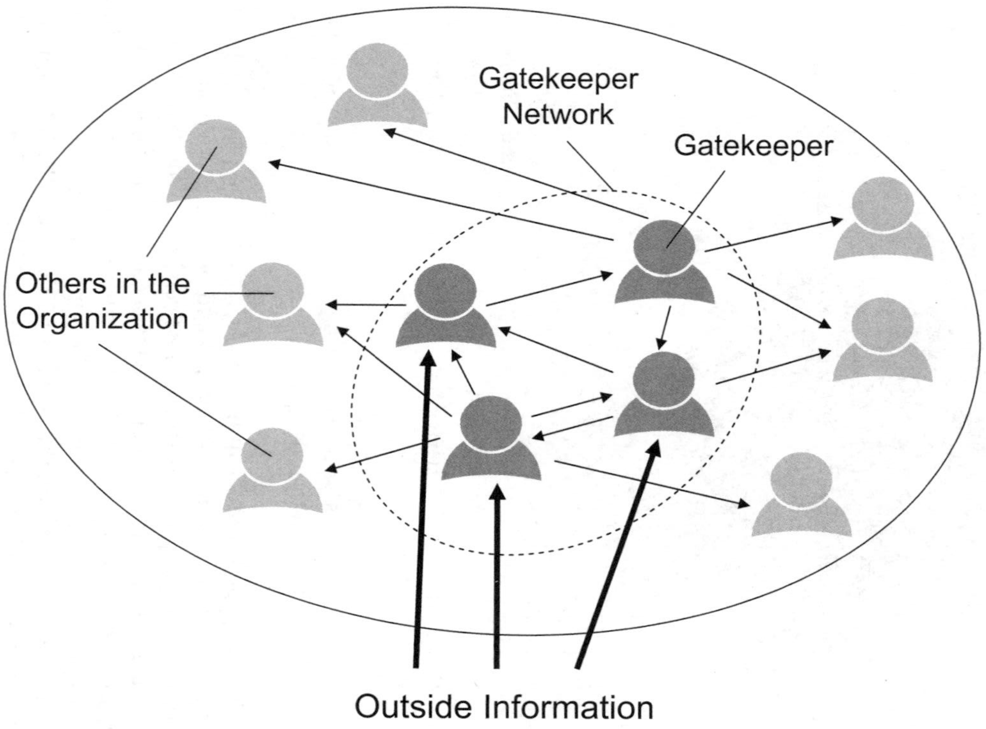

Figure 6.23 The function of the gatekeeper network and its interaction with outside information

Gatekeepers' tasks and communication

The gatekeeper is responsible for appropriate observation of her/his issue. Thus, they have to organize themselves, especially to define the information sources. However, other employees are also asked to provide them with relevant information, whenever available. Therefore, gatekeepers have to be known by the employees, and thus the gatekeeper network becomes formalized. Allen (1986: 161) is critical of facing a formalized gatekeeper network: this might even be counterproductive. However, there is no evidence for this anxiety. Myers (1983: 205) is even positively disposed to formalization, and numerous practical examples of a formalized gatekeeper network prove its eligibility.

The gatekeeper has to consider three main aspects: "technology," "market," and "competitors." It is a prior objective to keep track of any changes and developments regarding these aspects. Therefore, the gatekeepers are responsible for information collection, analysis, and communication by themselves. It is up the gatekeeper to define a detailed task in order to accomplish the prior objective, because they are supposed to be the people most capable of doing this. Hence, they should present the planned tasks to a Opportunity Landscape advisory board for approval. This board may consist of Executive Board members.

It is difficult to estimate the additional time exposure of a gatekeeper to their "normal" tasks. Basically, a gatekeeper keeps doing their job as usual, they just formalize knowledge and insight about their issue. To a certain extent, this could mean additional efforts for an individual gatekeeper, but in total, due to obvious redundancies, overall company efforts should be optimized.

The Opportunity Landscape relies on information push and information pull processes at the same time. On the one hand, the gatekeeper can present insight from observation by different proactive means: for example, regular meetings, reports, or anytime an important event happens. This means a certain documentation and management of the generated intelligence. On the other hand, the gatekeeper can be contacted anytime information about their topic is needed. Thus, the gatekeeper network, i.e. the Opportunity Landscape, is a corporate knowledge base for facts and trends from the company's environment that is accessible to anybody at anytime (when the gatekeeper is accessible). Because every gatekeeper becomes a kind of irreplaceable employee, a knowledge preservation strategy has to be defined. However, mixing a codification strategy with a personalization strategy is not optimal because executives who try to excel at both strategies risk failing at both.

All gatekeepers together build the gatekeeper network, which can additionally be animated and coached by a coordinator. The coordinator should not be considered a technical supervisor of the gatekeepers' work, nor as central communication node. They are another participant in the gatekeeper network with another focus for their contribution. For example, they provide supporting tools, i.e. methods and infrastructure. Typically, the coordinator is a member of the technology management group. (See Figure 6.24.)

Taking action from insight and updating the Opportunity Landscape

From the perspective of communication logic, taking action from the Opportunity Landscape happens with a push and pull logic. As we have already seen, the *first* impact of the Opportunity Landscape for innovations is to give input to the screening process. This is a proactive information push in the case where a gatekeeper identifies an action requirement on an insight from her/his issue.

Also with regard to the screening process, any gatekeeper can be contacted for expertise during a screening project. A gatekeeper, for example, might become a member of a temporarily limited *ad hoc* screening team for further investigation. Such a "reactive" action is also possible with other business processes for any (decision-authorized) employee, whenever a decision in the innovation process has to be made, which is a *second* very important impact.

A *third* impact is the interaction with business mission and strategy (Figure 6.25). Since the Opportunity Landscape is also future oriented, knowledge from it may influence strategic decisions, even update business mission and strategy (here Technology Intelligence can be seen as a driver for a today-for-tomorrow strategy). In turn, a revised business mission and strategy may also influence the issues in the Opportunity Landscape, i.e. when a new technology field is to be explored. Thus, the Opportunity Landscape becomes updated. Another update input, however, comes from gatekeepers themselves. Since, they are the experts with intensive contact to external experts and opinion leaders (i.e. with the ITI), they are most competent to decide which issue might be added and which existing issues are becoming obsolete, or to decide simply to change the depth of observation. Thus, the Opportunity Landscape is a dynamic process. The dynamic nature of this process and the gatekeeper network infrastructure support knowledge management and organizational

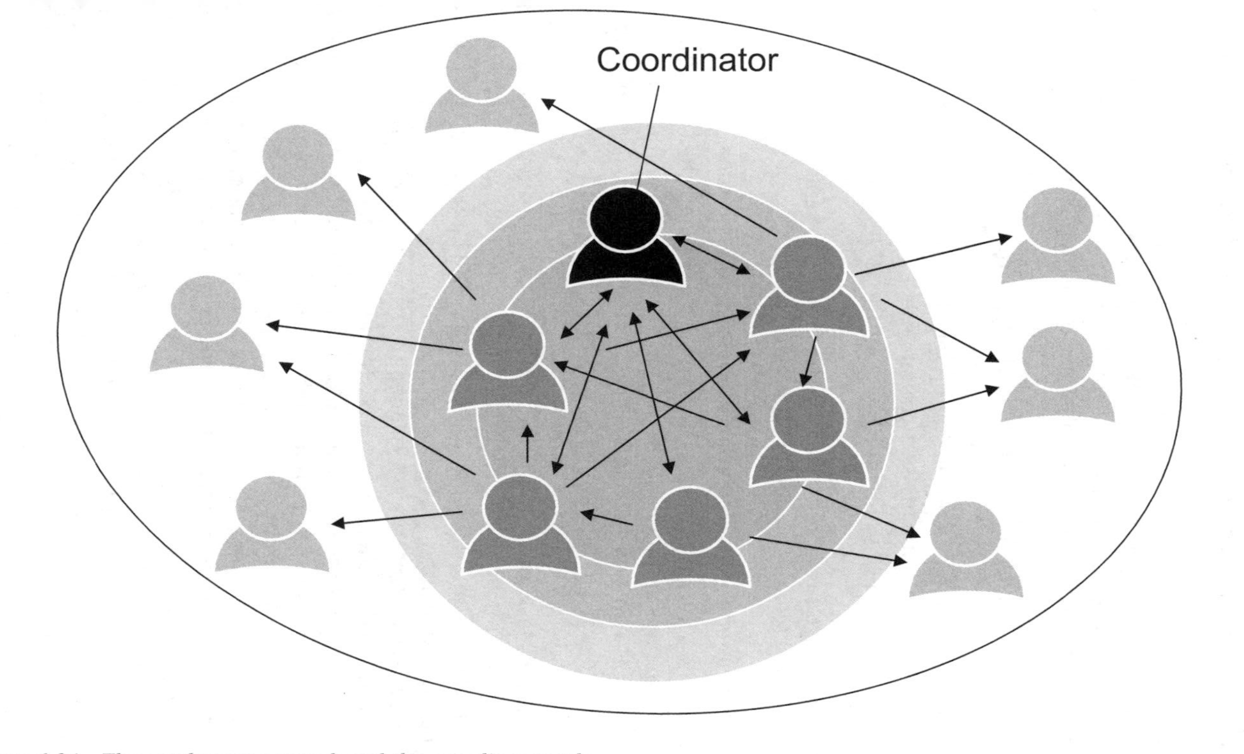

Figure 6.24 The gatekeeper network and the coordinator role

learning, respectively, by deliberately building up the knowledge base. This knowledge base can be used to develop new internal competencies.

A *fourth* impact is proactive information delivery from a gatekeeper when an alarming event occurs. Similar to the input into the screening process, the gatekeeper has to assume the responsibility of discharging this information to the concerned person, which is in most cases top management.

All these impacts combined make the Opportunity Landscape:

- a knowledge base concerning present and future facts and trends that form the company's environment,
- an alert system for discontinuities in the company's environment and
- a proactive input generator for the fuzzy front-end of innovation i.e. the screening process at Straumann.

Implementation of the Opportunity Landscape at Straumann

Unlike the screening process, the implementation of the Opportunity Landscape was an independent project. The project team consisted of a core team with two members of the technology management group and a larger group including the Executive Board and senior management (heads of R&D, Marketing & Sales, Finance, and Production). The core team's task was to plan and moderate the implementation project as well as further develop any details, i.e. the gatekeepers' task specifications and the reporting contents. The larger group met three times for workshops.

The **first workshop** was designed to define the strategic fields and the issues within them. It was very important to have a multidisciplinary team to fill the Opportunity Landscape with its contents, which in fact reflect the entire company's future. However, since all of the participants were from the company, there was a lack of an external and objective point of view. During this workshop, the top-down approach was chosen first. The business mission did not allow for deriving strategic fields though, because it had been too narrowly defined, and there was a danger of ignoring some important issues from outside the mission statements. Therefore, the bottom-up approach with the brain-writing method was chosen: about 120 issues were written down, consolidated, and grouped. Finally, a total of 8 groups became the strategic fields and were compared to the business mission.

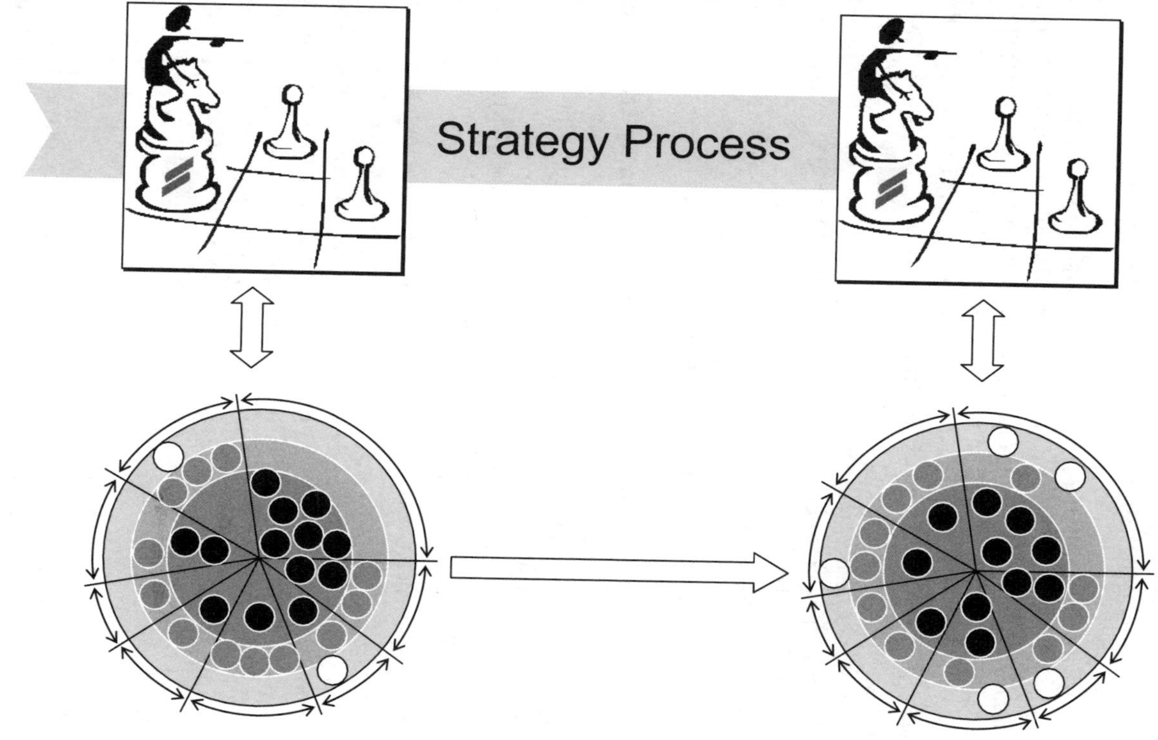

Figure 6.25 Interaction between the Opportunity Landscape and business mission and strategy

The **second workshop** was held to define one gatekeeper for each issue (the issues had been approved by the Board of Directors). It turned out that some employees became gatekeepers for more than one issue, while some external experts had to be assigned to gatekeeper functions.

The **third workshop** defined the gatekeepers' job specifications and the reports' contents. In addition, the roll-out of the Opportunity Landscape was planned, i.e., training and IT support for the gatekeepers. A member of the core team was selected as gatekeeper coordinator to fulfill training tasks and to support the gatekeepers with analytical tools. This coordinator plays an important role in the gatekeeper network: they are informed to the maximum extent about insights from each gatekeeper, and they should be able to assure an optimal information flow between them. The result should be that any redundancies are avoided.

These workshops took place every two months. With an additional preparation time of another two months before the first workshop, the implementation phase of the Opportunity Landscape (up to the pilot phase) took a total of six months.

To summarize, the implementation of the Opportunity Landscape followed rather a classic project management design. According to the screening implementation, the implementation of the Opportunity Landscape was run in parallel with the design process (after first conceptual reflections). The various workshops showed that Opportunity Landscape implementation was also a highly participative process.

Practical experience with the Opportunity Landscape at Straumann

The Opportunity Landscape is a novel concept that has been developed within the company. Thus, there was no prior experience with this concept in another enterprise, nor had any reflection been made on this issue at Straumann. Therefore, design and implementation of the Opportunity Landscape was rather explorative. It is not surprising that actual experience did not fully realize all our expectations. In this section, the facts are described first. Then, suggestions for improvement are presented.

Facts

During the workshops, 27 **issues** were defined: 13 in the player area, 12 in the substitute area, and 2 in the junior area (cf. Figure 6.22). This is an imbalance with a focus on "today issues." The presence of only two issues in the junior area is critical, because if juniors are missing today, player issues will be missing tomorrow. A gatekeeper could be attributed to almost all issues. As described in the last chapter, some employees became gatekeepers for more than one issue, and for some issues no internal person could be found. Experience over about one year was disillusioning: about one-third of issues presented results that corresponded to the gatekeepers' tasks, one-third of the issues were "on hold" because the attributed gatekeepers were on sabbatical for several months, or they left the company in the meantime, and the last third of issues simply were not completed. Thus, the Opportunity Landscape did not work as expected after one year.

However, some **action** was taken from insight. A constant interaction between the (performing) gatekeepers and screening core team members accentuates the synergetic potential of these two Technology Intelligence elements. Also, a proactive behavior is important in order to deliver interesting information to other employees and top management. Another action was the adaptation of business mission and strategy that might have been provoked by insights from the Opportunity Landscape.

Other interesting facets are **time budget** and **finances**. With regard to time budget, drawing a clear line between gatekeeper activities and "normal" work time is delicate. An estimation of time spent by the (performing) gatekeepers was 10% of their working hours. This would mean 2.7 full-time jobs for 27 issues. But considering the fact that a gatekeeper has already been involved in his field with some information-collecting activities, and that prior redundancies should be avoided, the effective additional time spend due to the Opportunity Landscape is certainly less than 2.7 full-time jobs, perhaps even equal to zero. The coordinator function needed extra time during design and implementation, but finally tended towards zero. Financially speaking, the budget for external costs consisted of traveling costs for congresses, etc., database and journal registration, and some IT support. In fact, since most of these costs were already budgeted for the business functions, a simple shift to a centralized Opportunity Landscape budget took place. The amount for this was budgeted at US$150,000.

Suggestions for improvement

These facts have shown that the state of the Opportunity Landscape is not exactly where it was supposed to be. But all things considered, the Opportunity Landscape nevertheless is deemed to be a promising approach, "to lead to effective decision-making for future innovations by building a knowledge base about present and future facts and trends." Experience gained during about one year at Straumann suggested two major paths for improvement.

A reason that the Opportunity Landscape has not been fully accepted yet might be the fact that the business mission and strategy did not allow for giving the Opportunity Landscape enough weight. This means that the Opportunity Landscape activities were not considered high priority by the gatekeepers (those who did not do anything), or by top management (no pressure on the gatekeepers and no substitution of issues "on hold"). This is in opposition to top management's commitment that "the Opportunity Landscape is exactly what we needed." This behavior can be explained by Argyris et al.'s (1987: 83) "theories of action": what people do often differs from the theories (or strategies) they espouse. Thus, a suggestion for improvement is to scale down the Opportunity Landscape to a defined unit, like the R&D function, with a small gatekeeper community functioning on a voluntary basis (but approved by management) and with a clear client (for example, the head of the Technology Center). This would create success stories and could have a catalytic effect on other units and the entire company.

A lot of responsibility was given to the gatekeepers. They should have defined their tasks in detail and how they should be accomplished, because they were supposed to be the most competent to do this. Obviously this did not work for all gatekeepers. A suggestion for improvement is that the coordinator should train and discuss the content with each gatekeeper individually. She/he should support the gatekeeper with a definition of the scope of their issue as well as with the possible tools. Through these individual meetings gatekeepers could also express their objections without demotivating other gatekeepers, or without losing their credibility.

IT Platform

The IT platform is briefly described in this chapter, but I have avoided delving into technical details. The benefit of a common IT platform is twofold: decentralized data storage and communication support. Since several people have to revert to the same documents, it was necessary to decentralize available data in electronic form. Straumann already used some Lotus Notes databases, therefore this platform was also adapted for the TI system.

For screening, the **screening monitor** was defined (Figure 6.26). In fact, the screening monitor is just a data storage tool that differs only slightly from a simple file server solution. However, graphically speaking, this tool allows a more attractive and more comprehensive working environment. In addition, the screening monitor assists the screening core team on-line during the monthly meeting. This tool is the database in which any screening input (accepted or refused) is stored. Therefore, the history of any screening case can be reconstructed.

Analogous to the screening monitor, a Lotus Notes database was set up for the **Opportunity Landscape** (Figure 6.27). This is a knowledge navigator within the gatekeeper network. Most important is a who-is-who representation for identification of the gatekeepers. This tool is not understood as a complete database for all documents that a gatekeeper has. The goal is to stimulate direct contact with gatekeepers in order to increase informal direct communication.

A new Technology Intelligence system at Straumann

The goal of this chapter is to discuss the generated Technology Intelligence elements at Straumann as compared to the initial situation at Straumann. We have also referred to the theory described in Chapter 4. Therefore, the discussion again follows the framework of the Technology Intelligence value chain. Particular attention has been given to avoiding repeating statements that already have been formulated in previous chapters. However, some information redundancies appeared to be inevitable.

TI management

Since Technology Intelligence has been formalized at Straumann the TI system has direction. How the system was designed, directed, and developed, as well as implemented, was the purpose of this action research case. It has particularly been shown that the system and its elements were not conclusively designed in advance, and then implemented. The TI system is rather subject to a dynamic learning and adaptation process. Therefore, next to the TI process, the TI system process itself contributes to organizational learning.

The crucial roles in this dynamic management process are occupied by "champions." The team, consisting of an expert champion (the screening manager and Opportunity Landscape coordinator) and a political champion (head of Technology Center), promises to sustain a steady adaptation and improvement of the TI system. The Technology Intelligence manager's task is to manage the TI system and to coordinate elements and activities, not to fulfill TI activities. However, he is the gatekeeper of one issue and he is actively involved in the screening process.

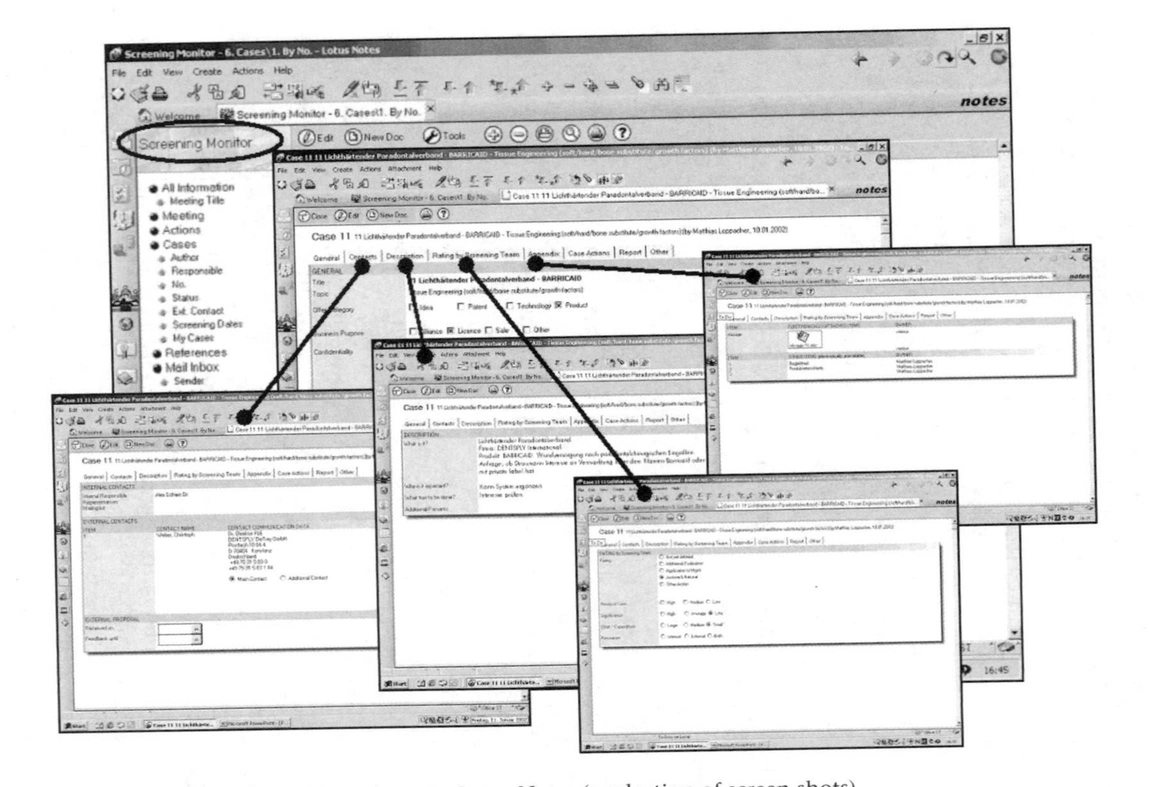

Figure 6.26 The screening monitor at Straumann in Lotus Notes (a selection of screen shots)

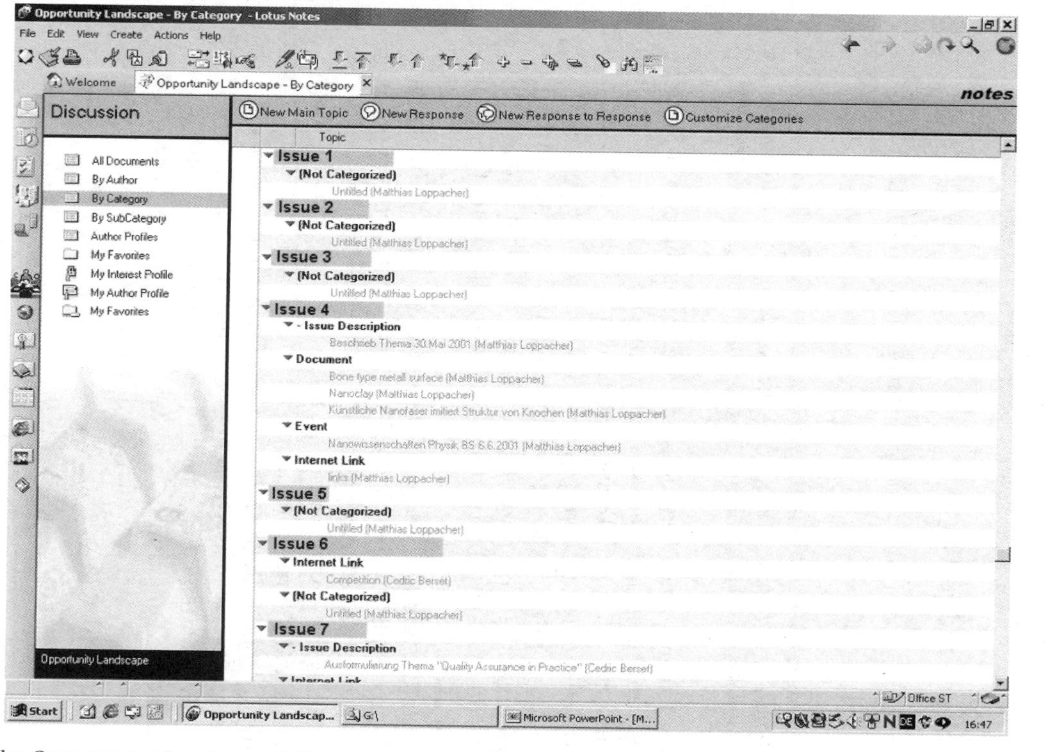

Figure 6.27 The Opportunity Landscape at Straumann in Lotus Notes (screen shot)

The dynamic TI management process needs a mission and goals in order to follow a defined framework. At Straumann, the TI mission and goals were important to set the scope of the TI system. Most notably the design and implementation of the Opportunity Landscape was a consequence of adapted TI mission and goals.

TI mission and goals

Two missions were derived from the business and Technology Center mission:

> (1) *to lead to effective and efficient decision-making before product development;* (2) *to lead to effective decision-making for future innovations by building a knowledge base of present and future facts and trends.*

The elements that were designed and implemented to accomplish these missions are the screening process and the Opportunity Landscape. The goals of these two elements are shown in Figure 6.13.

The TI mission and goals at Straumann are consistent with reflections in the existing literature – especially the fact that the TI mission and goals are related to business mission and strategy, and reflections about observation foci are in line with the solution at Straumann.

Insight could also be gained into strategy itself. While Straumann does not explicitly have a "dual strategy," this approach could nevertheless be observed in relation to TI. While the "today-for-today" strategy builds the field of action, i.e. the decision criteria, of the screening process and defines some issues of the Opportunity Landscape, the latter also builds the basis for new business strategy. Therefore, strategy certainly drives today's TI activities. But in addition to this, TI drives tomorrow's "today-for-today" strategy. And in fact, since Straumann installed the Opportunity Landscape, they implicitly have a "today-for-tomorrow" strategy, which is closely related to the content of the Opportunity Landscape.

TI structures

The Straumann TI structures discussed above are still valid within the new TI system. The difference between this initial situation and the new TI system are two adaptations. Firstly, two new structural elements were introduced: technology management and the gatekeeper network within the Opportunity Landscape. In addition, the screening element was reorganized and personalized. Secondly, the TI activities are now formalized. The structural organization of the TI system at Straumann is illustrated in Figure 6.28.

The main characteristics of this structural organization are:

- **Decentralized TI workers**: Because of restricted resources in this SME, TI activities are shared among employees as widely as possible. Attention is given to allocating activities depending on competencies and on available time.
- **Internal and external networks**: In order to profit from best and direct information sources on the one hand, and in order to transfer and make the generated intelligence available to most decision-makers ("organizational intelligence") on the other hand, the new TI system is based maximally on a network logic.
- **TI coordination**: The TI system is actively managed from the technology management group in order to coordinate activities and to avoid redundancies.
- **Continuous and discontinuous elements**: Next to continuous activities (for example Opportunity Landscape, screening core team, etc.), the TI system also allows discontinuous activities (for example screening investigations, ITI projects, etc.).

Several roles in the TI system are of particular interest. First of all, the **TI coordinator** is a screening manager and a core team member, an Opportunity Landscape coordinator, a gatekeeper, and a TI champion all in one person. Compared to roles in the existing literature, the TI coordinator at Straumann takes on the roles of the expert, process coordinator and method specialist. Secondly, **screening core team members**, **screening investigation project members**, and **gatekeepers** take the roles of scanners, information specialists, and analysts. In fact, they are the value-adding persons within the TI process. Thirdly, regarding participants in a **informal network** (i.e. internal and external experts), they may take the same role as people within a formal network. Lastly, the head of the Technology Center is the facilitator or the political champion of the TI system.

Consequently, the roles that have been observed and introduced at Straumann cope very well with roles described in literature. In this case, a difference between roles at Straumann and roles in large companies cannot be discerned, which means that apparently resource restriction does not change the roles on their own but change the content and scope of the same roles.

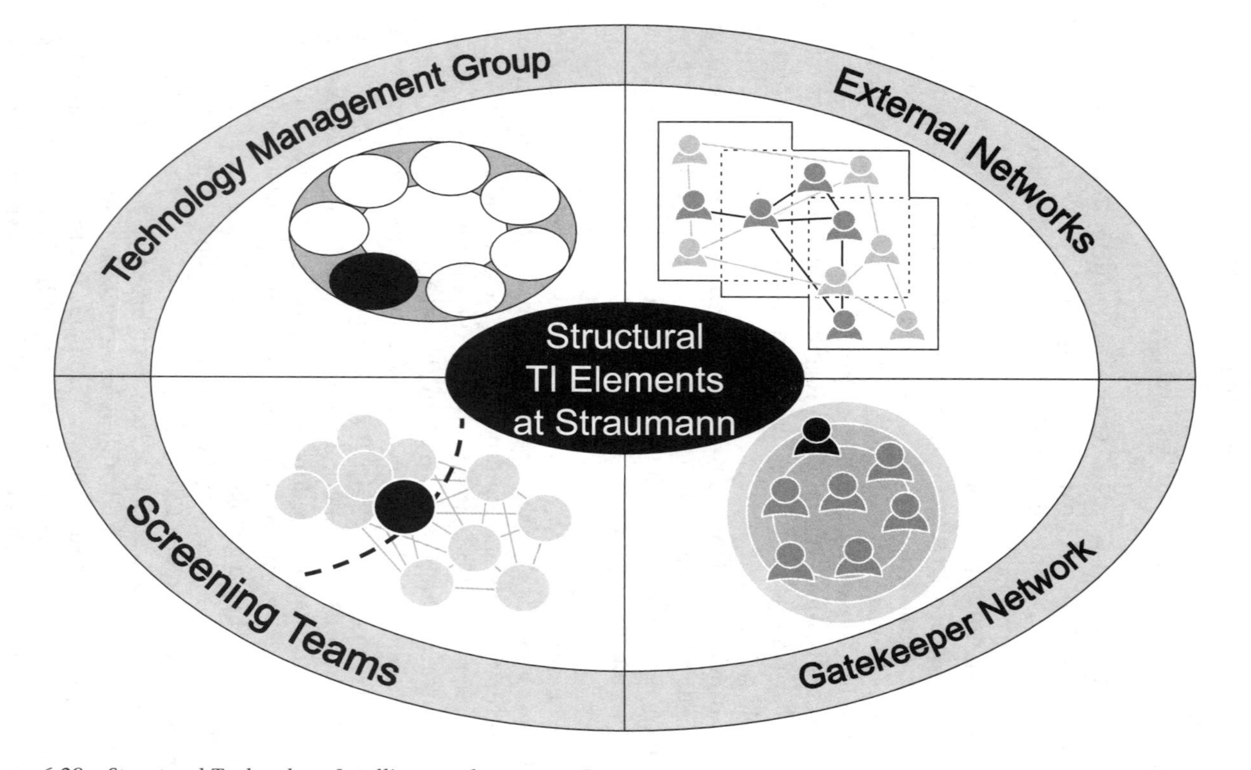

Figure 6.28 Structural Technology Intelligence elements at Straumann

The TI process

The main TI processes at Straumann, the screening process and the Opportunity Landscape, have already been described in detail. The information will not be repeated in this chapter. Alternatively, in order to be comparable to the existing literature, this chapter contains a short discussion of activities in the idealized TI process.

Formulation of information needs

The TI mission and goals provide a first rough framework for TI activities (Figure 6.13). This framework gives an idea about the time focus and observation areas. By this, information need is formulated explicitly (for example, within screening) and implicitly (for example, by asking gatekeepers for advice). At the same time, TI activities are pursued reactively (for example, through a screening investigation project) and proactively (for example, gatekeepers have to define their observation scope and sources). Thus, referring to Figure 4.9, both observation perspectives ("inside-out" and "outside-in"), and both impulses for TI activities (reactive and proactive) are considered within the TI system at Straumann. In addition, a permanent change of emphasis can be observed. This is even the nature of the Opportunity Landscape as a knowledge-base of facts and trends: gatekeepers have the responsibility to remain informed within their fields (i.e. they have to seek proactively for trends from anywhere), but at the same time they have to respond to inquiries on demand. Thus, their role permanently changes.

Information collection

Roles of employees who collect information have been described before. Again, information collection activities are shared as far as possible by diverse employees depending on competencies and on available time. Actively involved in information collection are gatekeepers, the screening core team, and investigation project members. However, other employees and external experts, in particular ITI members, can be considered as passive information collectors.

With regard to information sources, there are almost no differences from the initial situation at Straumann. The exceptions are some additional links that gatekeepers had to define. Thus, some informal information sources were made more transparent.

The experience at Straumann showed that access to information sources (formal and informal) does not cause major problems. Therefore SMEs are, *a priori*, not at a disadvantage in terms of access to information sources when compared to large organizations. This allows a presumption that the challenge of Technology Intelligence is not to have access to information, but to convert information into intelligence that is usable for the company, and which is made usable through analysis.

Information analysis

Information filtering is the core business of the screening process, in particular to improve efficiency and to avoid redundancies. In the Opportunity Landscape, filtering happens on an individual level. Each gatekeeper has to organize herself or himself in order to avoid information overflow.

Information is integrated into the company and decision-making processes by means of participation of decision-makers in the TI process (for example, screening and the Opportunity Landscape). If relevant information is of interest to people who are not involved (for example, the board of directors), direct communication is favored.

Information assessment at Straumann is based on optimal expertise. Three examples are the screening core team, screening investigation projects, and gatekeepers. The screening core team is multidisciplinary in order to have a holistic view. Thus, any screening input is judged with a holistic view. This is also "optimal" expertise because this process is fast and does, therefore, not waste resources. Then, screening investigation project teams are composed of internal and/or external experts in order to achieve best analysis results. This is a very flexible form for optimizing expertise. Finally, it is the nature of gatekeepers to be experts within their issue. It is up to him to complete competencies for information analysis by means of other internal or external experts. So far, Straumann does not utilize technical analysis tools.

Information dissemination

As mentioned before, the main dissemination channel is participation. For example, the presence of a chief executive during screening meetings (the monthly meeting *and* the screening lunch) assures direct communication with top management. If participation is not possible, i.e. in gatekeeper work, direct and oral communication is preferred. Transparency on "who knows what" seems to be very favorable for communication. This is particularly assured in the Opportunity Landscape.

On the whole, communication follows a combined information push and pull logic. This logic is well described above and will not be repeated here.

A key role to animate and direct communication certainly is taken by the TI coordinator. In fact, he is involved in all processes and is supposed to have the best overview of TI activities, and therefore can straighten out communication barriers. However, he should be a mediator rather than another communication node.

Information application

Two key aspects of information (or intelligence) application can be considered: use for decision-making and learning.

Management decisions are always made when a screening input is judged. Therefore, intelligence workers and intelligence users are the same persons in the screening process. However, sometimes a screening decision has to be made by top management, for example because investigation costs are high. In this case, decisions should be interactive, which means that top management should avoid simply making yes-or-no decisions, but also justify their decisions. The experience at Straumann unfortunately did not fit this ideal vision. However, since the Opportunity Landscape is the organizational knowledge-base for facts and trends, anyone may ask for the gatekeepers' advice whenever a decision (for example, a decision within a development project) has to be taken. Therefore, the TI system becomes a veritable organizational intelligence system.

In addition, since all TI processes are participative, organizational learning is fostered. Since the TI system is not simply a private, top management support affair but is directed at all employees, the investment in the TI system seems to be justified.

TI tools (methods and infrastructure)

The "Straumann 1999" case study indicated a wish for a more systematic and coordinated use of both qualitative and quantitative TI methods. But experience during the design and implementation of the TI system at Straumann, in particular during workshops, showed a certain suspicion and, therefore, refusal of technical TI methods (for example, bibliometrics, S-curve analysis, learning curves, etc.). The use of such methods does not seem to be in line with company culture, which instead is based on multidisciplinarity and intersubjective opinion forming. Therefore, emphasis has been given to support this culture, which resulted, for example, in the screening core team and the gatekeeper network approaches.

In contrast, the set-up of the IT platform for the TI system (screening monitor and Opportunity Landscape) underlines the acceptance of the technical infrastructure. The screening monitor, for instance, is used during the screening meetings.

Improvement in decision-making (TI measurement)

In-depth insight into the effectiveness and efficiency of the TI system has already been given in the chapters about implementation and practical experience through the screening process and the Opportunity Landscape.

A formalized TI performance measurement system or indicator was never a real topic, nor a precondition to introduce the TI system at Straumann. Many reasons for this are imaginable. Firstly, company culture was open to simply giving the TI system a try. Secondly, Straumann could afford this try because of its favorable financial situation. Thirdly, the need for and the value of a TI system was estimated intuitively by top management. All in all, the absence of a detailed TI measurement gives the TI system a certain degree of freedom and does not waste resources for control. But a danger of this is the fact, that the TI system can easily be brought down, for example by an important change in top management, or if the financial situation forces the company to focus on direct value adding activities.

Even though measurement was not a must, several facts were examined by the TI coordinator. One fact was described in the practical experience with the screening process: Five project proposals emerged from 72 screening inputs during about one year. This could be defined as an indicator to be compared with experience in the following year. Another indicator for measuring efficiency could be the number of inputs in the waiting position (21 of 72 inputs). With regard to the Opportunity Landscape, an alarming sign was the fact that just one-third of gatekeepers fulfilled their tasks. This can be interpreted to mean that Straumann only captures one-third of facts and trends that are relevant to its future! Another TI measurement opportunity was benchmarking with other companies. In fact, this is the content of the next main chapter about further validation of the generated TI solution.

Another point of interest is comparing the budget for Technology Intelligence at Straumann with amounts suggested in the existing literature. In fact, Straumann spent about US$162,000 for running costs, and about US$12,000 for IT investment, and utilized about 3.2 people (= 6,000 hours/year) for the screening process and the Opportunity Landscape together. However, this does not take into account the fact that time can be gained with the Opportunity Landscape. Therefore this number must be handled with care. The existing literature gives a range of US$66,500 to $183,750 for running costs, $8,750 to $17,500 for software investment, and a minimum of 2,300 hours/year of manpower. Except for manpower, costs at Straumann correspond to those mentioned in the literature. Manpower at Straumann goes beyond what the literature suggests. However, the significance of the amount of 3.2 people is not guaranteed.

Conclusion: Technology Intelligence at Straumann

Conclusions will be drawn at two different levels: in the Straumann context and with regard to the general literature. These conclusions should be understood as tentative insight during a continuous process of learning more about Technology Intelligence in a technology-based SME.

Conclusions in the Straumann context

An in-depth case study on the initial state of Technology Intelligence at Straumann showed several gaps which were reformulated in terms of action requirements. Therefore, the new Technology Intelligence solution will be discussed with regard to these action requirements.

The **Technology Intelligence framework** results from several initiatives. TI missions and goals were defined based on business mission and strategy. Since business mission and strategy and the TI system are interactive, the TI system (consisting of several processes) becomes a business process. These TI missions and goals are the foundation of the TI system, which consists of six main elements. The two core elements are the **reorganized screening process** and the **Opportunity Landscape**. They are basically supported by an external expert network (i.e. the ITI), an IT platform and a coordinating function, which is the technology management group. All these elements together complete the **Technology Intelligence System** within the defined TI framework. Because this TI system supports activities all over the company and is, thus, a part of the general business process, and because Technology Intelligence is considered a singular task among others to be coordinated by the technology management group, there is a superordinated **Integrated Technology Management context**. Altogether, these initiatives let us conclude that the defined action requirements are conceptually completed. The TI system was a continuous process of implementation at Straumann and initial experience could be gained. This experience showed strengths and weaknesses of this system, which are described in the discussion of each element of the system.

The value added with the implementation of the TI system entailed various aspects. This system helps Straumann to accomplish its R&D mission, which is "to keep, increase and transform the innovation power of Straumann into market-oriented and profitable product concepts." Because the TI system design is based on a participatory approach, organizational learning is enhanced, which enables employees at any company level to contribute to the R&D mission. At the same time, since the TI system is not a "one man show" but a task in which the efforts of numerous people (in total at least 35 people are continuously involved: one TI manager, the head of the Technology Center, 6 screening core team members, and 27 gatekeepers, in addition to people periodically involved in screening investigation and informal networks), the fact that resources are restricted is taken into account as much as possible. Again, it is difficult to balance the additional resources used for the TI system with resource wasting avoided by the TI system.

The TI system as it is described in this chapter is not conclusive. Management of this system is an ongoing process of further development and adaptation. We could see that within both core processes some adaptations are necessary. The Opportunity Landscape in particular needs some adaptations to be a sound knowledge-base for facts and trends. However, this action research case study gave us a "longitudinal" view of the design and implementation of a TI system in a technology-based SME. The solution arrived at here is a reality adopted in a single company: Straumann. Therefore, the two main core elements, screening and the Opportunity Landscape, will be examined in two additional contexts, a matter we turn to in Chapter 7.

Conclusion with Regard to Literature

Several questions in relation to SMEs emerged during the discussion of Technology Intelligence literature. In this part, insight during the action research shall be compared to reflections done in the literature discussion. First of all, the discussion framework (Figure 4.2) seems to be appropriate for a Technology Intelligence system. Thus, this framework is again adopted for this discussion.

The longitudinal study of the design and implementation gave insight into the **management of a TI system**. In the Straumann case, the TI system has been formalized. This is not necessarily a precondition to pursue TI activities, because an informal system may also accomplish the desired missions and goals. However, management of a system seems to be clearer with a formalized system. There are four reasons for this: firstly, within a formal system, acceptance of TI activities is more promising than in the case of an informal system, because participants in the formal system are considered experts. Secondly, TI work is official work, which means that involved people can spend time on TI activities. Setting goals with management by objectives would be favorable. Thirdly, since transparency is a given in a formalized system, redundancies can be avoided. And lastly, for the same reason of transparency, the system is more likely to be coordinated through the three management functions (design, direction, and development) which were observed. However, in the early stages of the TI system at Straumann, the directing and the developing functions were merged. The Straumann case also demonstrates that starting from initial ideas, system design, and system implementation are synchronous processes with a permanent change of emphasis. This fast and synchronous procedure was possible because of the participation of most involved persons already during system design. This is probably an advantage of an SME compared to large organizations, because communication is very effective and efficient. In addition, the participation of most of the people to whom the TI system design may be relevant, and in particular the assistance of almost all executives during workshops, is probably not possible in larger organizations.

The action research underlined the importance of **linking the TI system with business mission and strategy**. At Straumann, the TI mission and goals were deduced from the business mission and strategy, which is an obvious link. With regard to the link between the TI system and business mission and strategy, a difference between Straumann and large organizations could not be observed. This lets us conclude that a TI system seems to be as important for SMEs as for larger organizations. The experience at Straumann also contradicted the objections of several authors that an SME cannot afford to run a formalized TI system because it lacks financial resources. However, since Straumann is perhaps at the upper limit of the SME definition, the solution generated here will be compared to start-ups in the next chapter. Considering the observation areas, the results of action research are in line with the literature. The two core TI processes, screening and the Opportunity Landscape, were based on the "keeping abreast" and "looking beyond" ideas, which correspond to the "inside-out" and "outside-in" philosophies. Straumann is committed to being a technological leader, and therefore "looks beyond." There is no insight into whether this area is reserved for leaders only, or whether followers also should act in this field (for example, to decrease production costs with a new process technology from outside the industry).

The **characteristics of TI structures** have already been summarized: TI activities are pursued in a decentralized manner and coordinated by a staff member. The system in general follows an "organizational intelligence" philosophy in order to take into account the fact that decisions can be made at any level in the company, and in order to enhance organizational learning. Also the existence of three types of TI activities (structured, informal, and hybrid coordination) could be observed. Factors that influenced the design of the TI system at Straumann were first of all innovation strategy, company culture, basic company structure, and available resources. Innovation strategy clearly positions Straumann as a technological leader with the mission to innovate incrementally and radically. Therefore, the scope of TI activities "looks beyond" Straumann's short-term business outlook. The organizational intelligence system, with formal, informal, multidisciplinary, and participative characteristics, copes with Straumann's open company culture. In fact, this company culture seems to be typical for the MedTech sector. The basic company structure influenced the structural organization of the TI system, and vice versa, in two ways. Firstly, the organization of the technology management group was a necessary adaptation of the organization of the Technology Center. Secondly, since the company is organized in line business functions, persons from each function were chosen in order to be most "democratic." One restriction on the design of the TI system at Straumann was the need to avoid additional personnel. Therefore, available resources influenced the design. As far as possible, all TI activities were shared among existing employees. One person, the TI coordinator, was employed additionally anyway, but he spends only 50% of his work time on TI coordination. The fact that resource availability influences the TI system design has not been highlighted in literature before, which again points to the focus of the existing TI literature on large organizations. Three factors that are mentioned in the literature – the technology life-cycle, the decision-making process, and the industrial sector – did not strongly influence the solution at Straumann. With regard to the technology life-cycle and industrial sector, Straumann is in an ambiguous situation; some technologies are emerging, some are mature, and Straumann acts in a highly scientific area as well as in the metal-processing industry. The decision-making process did not directly influence the system because at Straumann this process could not be clearly identified. Also, there was no evidence of intercompany or industry-wide cooperation in Technology Intelligence, and these are *a priori* interesting approaches. During the action research, the insight that TI is first of all an internal affair, even if the information gathered is external, has been confirmed more and more. Thus, industry-wide approaches seem to be valuable information sources, but cannot replace a TI system in a company.

Insights gained from the TI process at Straumann and a comparison to findings in the existing literature were the topics of the previous chapter. Here, another question will be considered: is the TI process really a process? As we guessed before, the idealistic representation of TI activities in a process does not reflect the companies' reality. In fact, experiences at Straumann are in line with the information extension (conveyance) and information reduction (convergence) character of Technology Intelligence. This means that TI is an ongoing process with an explorative character as well as a focused character. The exploration side is probably a long-term learning process in order to better understand the company's environment. The focused side tends to be oriented on short-term decision-making. With this view, the main question is not how the process should be organized, but how Technology Intelligence should be linked with processes in which decisions that need external information are made: for example, strategy development, resource allocation, portfolio management, etc. Action research at Straumann showed how the two core elements of the TI solution are linked with some other processes. While the screening process is a decision-making process on its own, the Opportunity Landscape builds the explorative knowledge-base.

The tendency against quantitative and in favor or qualitative **TI methods** described in literature could also be observed at Straumann. Acceptance of quantitative methods was low. The barrier to use of such methods is high because they are mostly time intensive and complicated. Knowledge of how to apply such methods, including interpretation of results, was not a competence of most employees. The question that arises is whether a simplification of methods would be helpful. We can describe three groups of methods. The first group includes methods that simply cannot be downsized. For example, a patent frequency analysis always remains a patent frequency analysis. The second group contains methods of which the depth may vary: an S-curve analysis, for example, might be based on estimations rather than on historically correct data. The third group consists of methods that do not need to be changed because they already are very flexible in use. All in all, quantitative method use would not be restricted to large organizations, but there is a gap between this theoretical possibility and reality. Qualitative methods, in turn, are used quite frequently.

In fact, the TI system at Straumann is entirely based on the use of such qualitative methods, for example intersubjective opinion forming, expert panels, expert interviews, etc. **Information technology** use at Straumann is restricted to storage and communication purposes. No intelligence specific tool – such as Data Mining software – is used. Therefore no insight into the usefulness of such IT tools for various TI activities in an SME could be gained.

The approximately two years of action research at Straumann, of which about one year consisted of practical experience with the TI system, yielded some insight into the **value added by improving decision-making**. This is extensively discussed in the practical experience parts of the screening and Opportunity Landscape description and in the new TI system description. In summary, the two most valuable contributions of the TI system to the company are improved basics for any decisions and enhanced organizational learning. Both aspects have contributed to the company's success. However, it is difficult to show a relationship between company success and TI system quality, because the TI system is only one component of decision-making, and only one way to push organizational learning. Therefore, a TI system should be considered as an indirect, supporting activity, like technology management, in the company's overall value chain. As mentioned before, the **costs of the Technology Intelligence system** at Straumann more or less match with the budget range that the existing literature indicates. Spending in terms of manpower, however, goes beyond those indications. However, since it is difficult to say how much time gatekeepers really spend in addition to their daily work, a human resources estimation remains rather indeterminate.

7
Further Validation of Generated Technology Intelligence Elements

The aim of this chapter is to discuss and validate, by means of cases, some elements of the generated solution for Technology Intelligence (TI) during the action research. Thus, this chapter does not aim to deliver evidence for the generated solution, but illustrates these elements in a context other than at the action research company.

Sample of validation cases

The validation framework consists of two dimensions (Figure 7.1). The first dimension is a **horizontal validation** of two generated Technology Intelligence elements in the context of other technology-based SMEs that are comparable to Straumann. The first element, the Opportunity Landscape, is discussed in the context of an instrument manufacturer of the Roche Diagnostics Division. The second element, the screening process, is compared to the screening process at Esec, a semiconductor packaging manufacturer. This horizontal validation aims, on the one hand, to illustrate that these elements are not unique to Straumann, which is the action research environment, and on the other hand, to explore and explain these elements in detail in order to build the basis for management principles.

The second dimension is a **vertical validation** of the overall solution of TI that was generated through action research. This vertical validation puts the generated solution in the context of technology-based companies that are very different from Straumann. Therefore, this vertical validation delivers additional insight into TI in use for companies beyond the initially defined company type. An initial comparison focuses on start-ups, which represent very small companies. The sample contains 14 Swiss start-ups. The second comparison is done with two very large organizations: Novartis and Daimler-Benz. This vertical validation builds another basis for management principles. Consequently four validation cases will be presented in this chapter.

Horizontal validation case A: Opportunity Landscape

After design and implementation at Straumann, the concept of the Opportunity Landscape was further developed and implemented at the Roche Instrument Center (RIC) in Rotkreuz, Switzerland, in the context of a superior technology management project.

Information in this chapter originates from seven workshops and numerous individual and group interviews during a time period of two months. For this project about 50 employees were involved in total.

After a short company description, the technology management project will be presented in order to illustrate the framework in which the Opportunity Landscape is integrated. Then, experience with the Opportunity Landscape at RIC will be discussed and compared with insight from Straumann. In order to avoid repetition, the conclusions of this section will be made in the overall conclusion of further validation of generated TI elements.

Roche Instrument Center (RIC): company profile

The Roche Instrument Center in Rotkreuz is a center of competence for instrumentation in biotechnology and in-vitro diagnostics (IVD). The RIC is the preferred supplier of innovative, high-quality, and low-cost instruments for the Roche Diagnostics business areas (Figure 7.2). RIC has about 550 employees.

The current business mission of RIC is:

> *We pioneer instrument solutions for continuous innovation of health information at the best customer value.*

The R&D process at RIC is mainly project-oriented (Figure 7.3). RIC is one partner among many within the Roche Diagnostics R&D process. At RIC, five main elements participate in the R&D process: R&D for centralized diagnostics, R&D for molecular diagnostics, the laboratory network system, R&D steering, and the technology group. The first two groups do project-specific instrument development, and together they include about 200 employees. Activities in these groups are coordinated by an R&D steering group that consists mainly of the heads of R&D, the CEO and the RIC CTO. The third group is an inte-

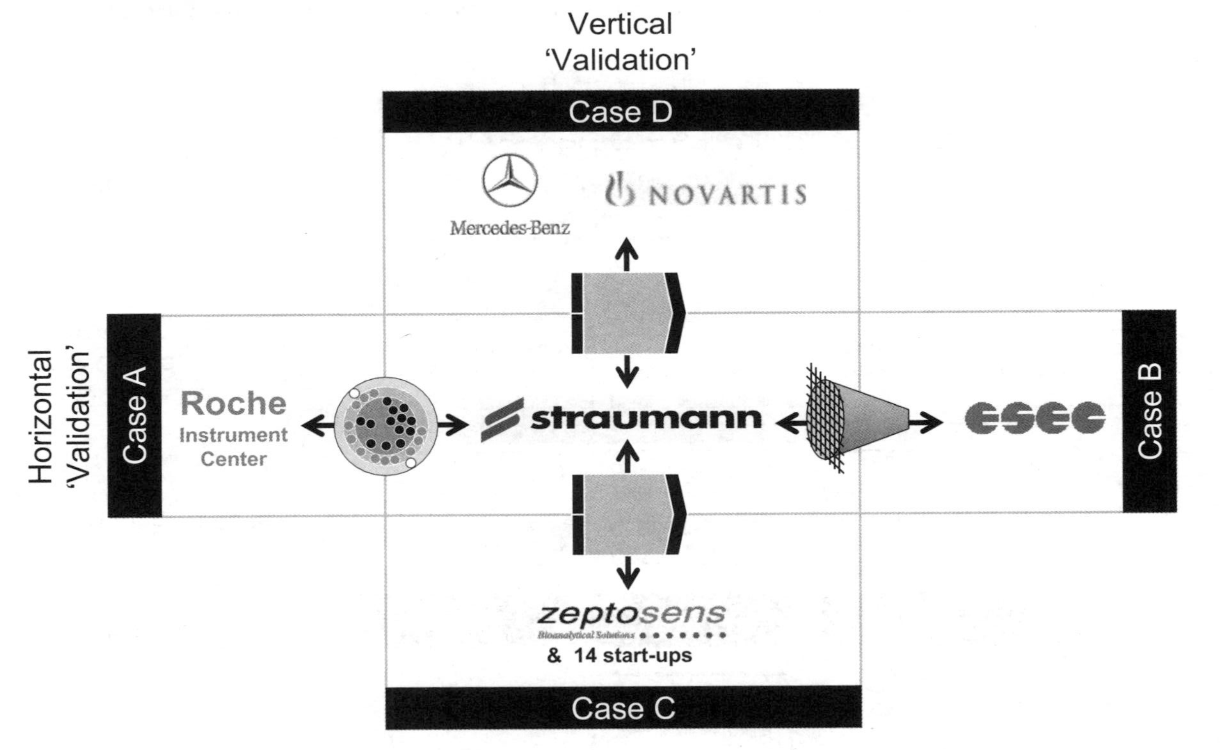

Figure 7.1 Horizontal and vertical "validation" of a solution for a Technology Intelligence system in an SME

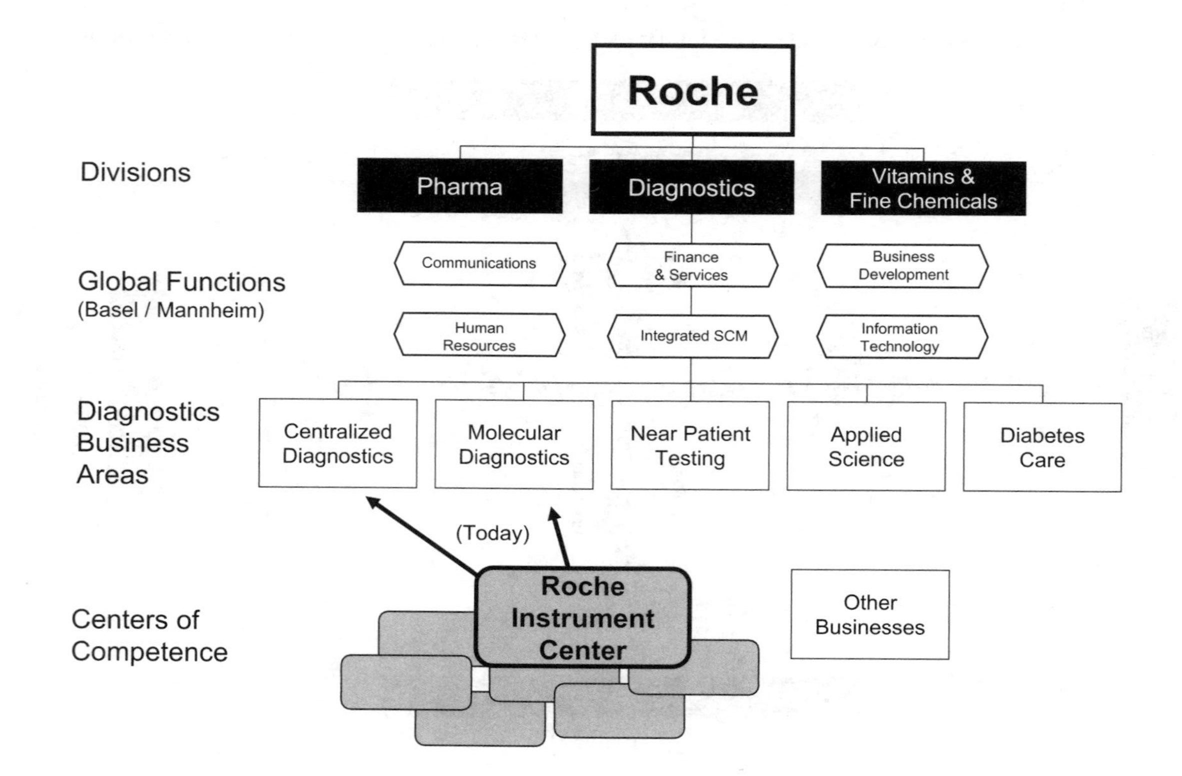

Figure 7.2 Integration of Roche Instrument Center in Roche Diagnostics

grative, non-project-specific solutions provider with about 5 employees. R&D groups and the laboratory network solutions provider are supported by a technology group, which is headed by the RIC CTO and includes technology specialists. This group performs basic technology development which may initiate a R&D project (input), as well as the group's work may be initiated by a R&D project in order to solve a technological problem (support). In addition, this group makes technology assessments for the R&D groups and for other units within Roche Diagnostics. R&D projects at RIC are coordinated by a project manager who is directly linked with Roche Diagnostics project managers.

Technology management at RIC

RIC's competitiveness today is based on the ability to adapt and utilize new technologies in order to develop instruments effectively and efficiently. Since the emergence of new technologies in the healthcare industry (HCI) is accelerating, business processes have to be adapted. In addition to this, the new business strategy at RIC evokes the need for a high level of competence in handling technological knowledge. Consequently, RIC decided to implement systematic and integrated technology management.

Coping with the following guiding principal questions is the task of technology management at RIC:

- Which technological competencies are necessary to maintain and improve competitiveness?
- Which technologies should be integrated in core modules, and how are these technologies utilized for products and systems?
- How much should be invested in technology development?
- How should technologies be acquired, internally and externally?
- When and how should a technology be introduced to the market, in terms of products or in terms of intellectual property?
- How should technologies and innovations be managed?

These guiding questions evoked three technology management elements that were implemented at RIC:

- Technology roadmapping (TRM)
- Technology Intelligence (TI)
- Technology platform management (TPM)

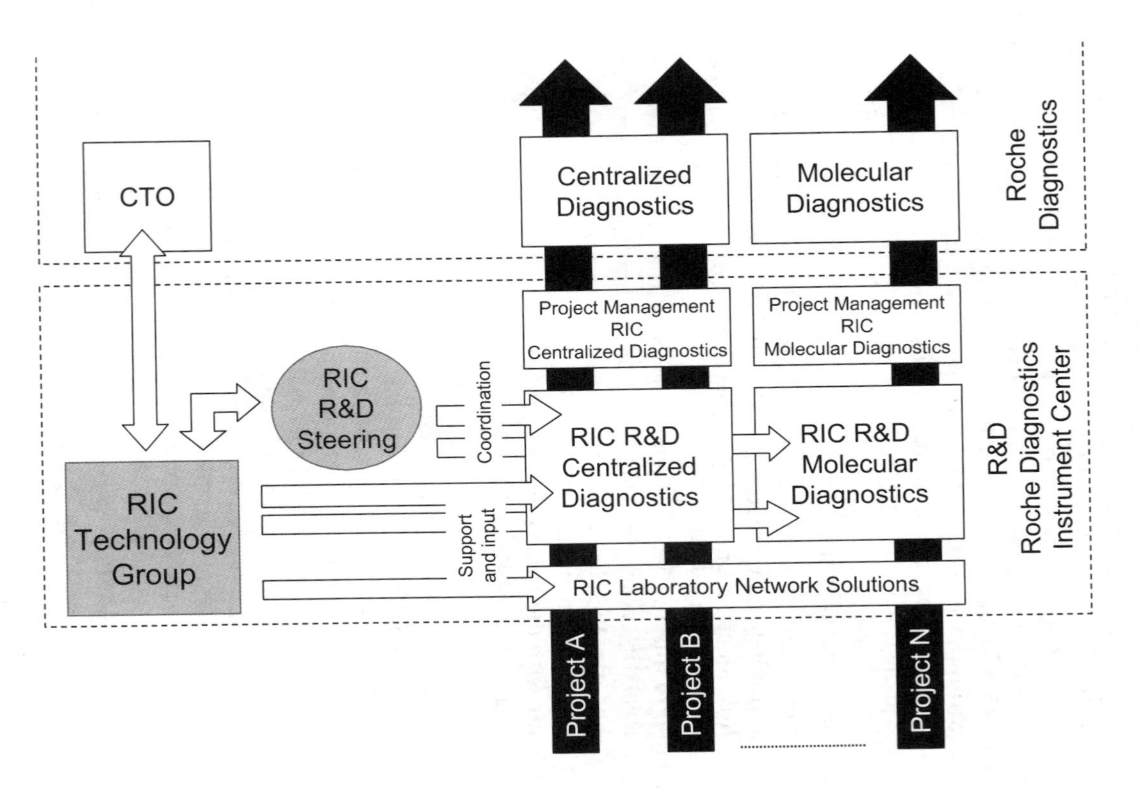

Figure 7.3 Organization of R&D at RIC

These three elements are complementary to existing elements at RIC: the Roche Diagnostics product roadmap and RIC technology policy and strategy. The interaction between all these elements is depicted in Figure 7.4.

In order to have a better understanding of the context in which TI is integrated, technology platform and technology roadmapping are explained briefly.

Technology platform management

Through a technology platform, RIC assures a long-term-oriented functionality that is independent from a Roche business unit or another client. To assure long-term-oriented means, in this context, to maintain and develop technological core competencies in order to bring them into line with the targeted core competencies. Therefore, within a technology platform (for example "analytical measuring technologies"), today's ready-to-use technologies (for example, "physical sensors") as well as necessary future technologies (for example, "reagents free detection) are managed. In total, six technology platforms assure the six core functionalities which can be combined for various RIC products, modules, and services (Figure 7.5). They are complemented by two basic technology platforms, i.e. general mechanics and electronics.

Each technology platform has its platform manager. Her/his main tasks are:

- to determine ready-to-use and future knowledge and technology domains within her/his platform and to update it regularly (circa twice a year). This task is particularly necessary after a new definition of targeted core competencies is determined (which means a reorientation of RIC), when the technology roadmapping process is completed, and if the technological environment changes dramatically.
- to coordinate gatekeepers in her/his technology platform.
- to select relevant information for technology roadmapping. Thus, she/he is the link between gatekeepers and the technology roadmapping team.
- to elaborate technology driven business ideas.

Technology roadmapping

RIC performs technology roadmapping in order to make business strategy operational in a systematic and step-by-step manner. This happens

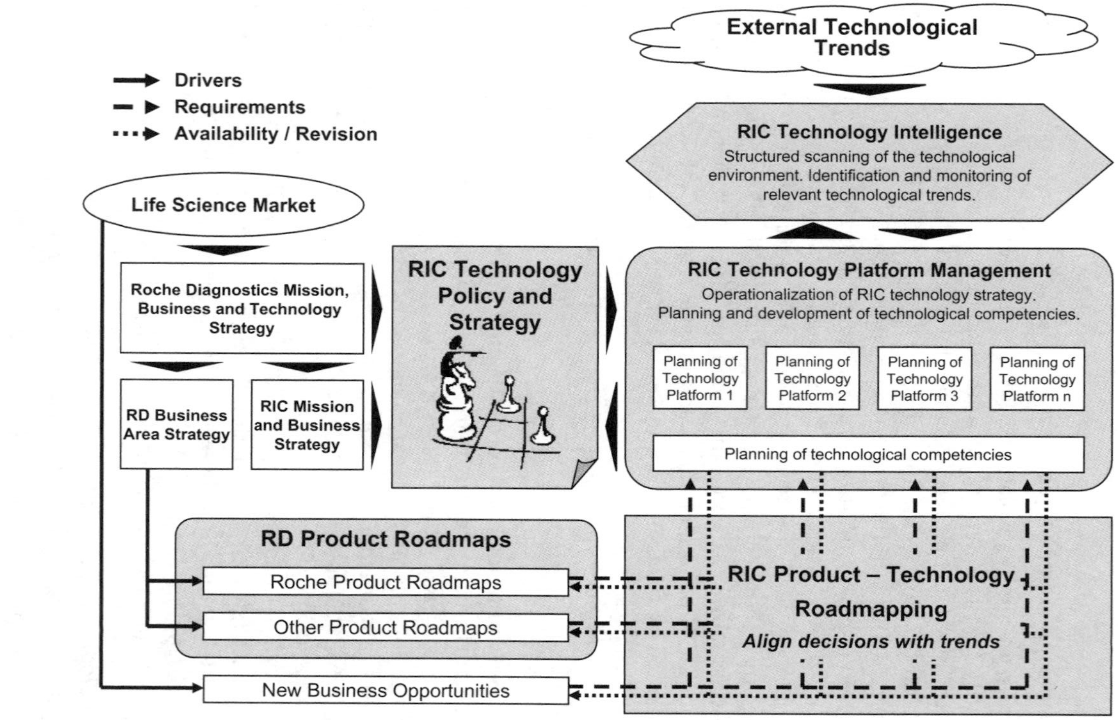

Figure 7.4 Technology management at RIC

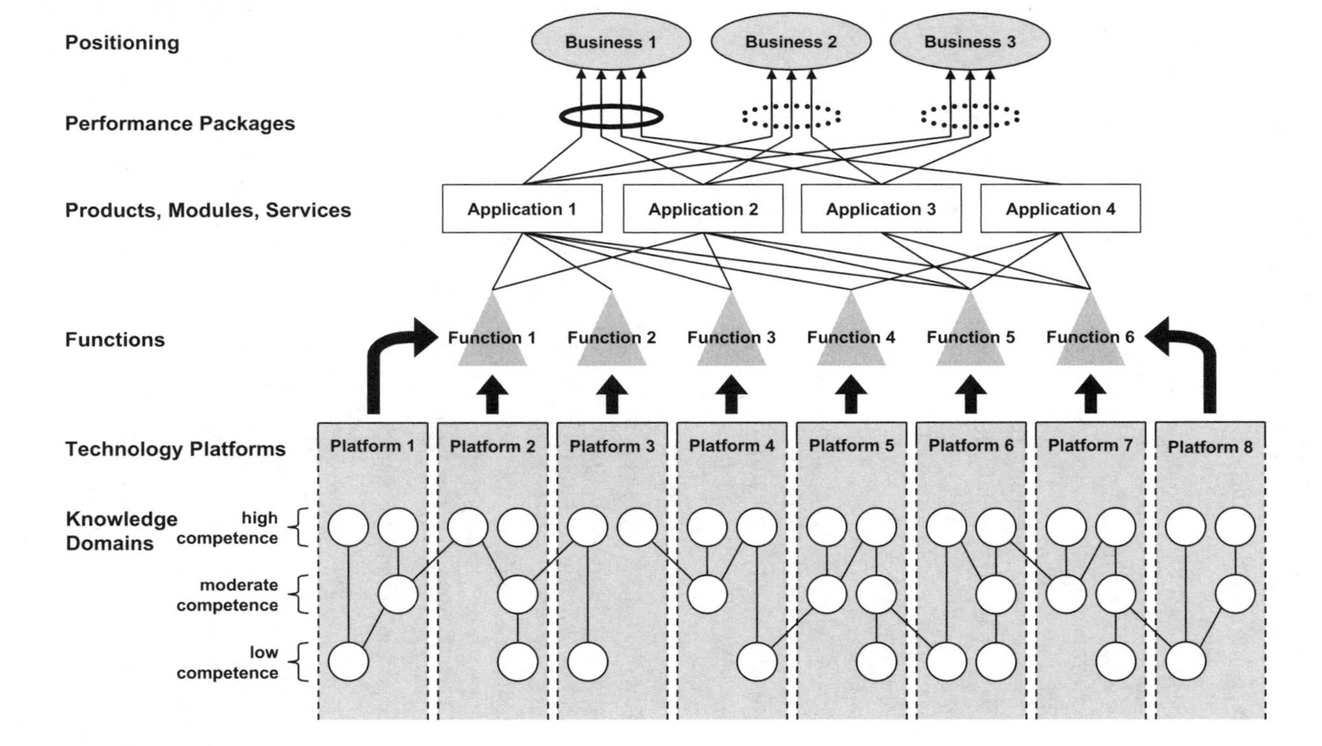

Figure 7.5 Function-based technology platform management at RIC

through matching internal competencies and resources with a forecasted future (by means of Technology Intelligence). Technology roadmapping generates and evaluates options systematically for action, identifies gaps of competencies, and then visualizes an optimal action plan, which is the technology roadmap (Figure 7.6). The technology roadmap is an integration of market drivers (know-why), products (know-what), technologies (know-how), and time (know-when). The time horizon typically covers two or a maximum of three product generations. This takes about 10 years at RIC.

Technology roadmapping is a process that is performed twice a year at RIC. This process includes six steps: (1) roadmapping team building, (2) collection of relevant information for roadmapping, (3) business option generation, (4) business option evaluation, (5) business option selection, (6) generation of technology roadmap. Participants in this process build the so-called technology roadmapping team. The members of this team are the head of technology, the CEO, the heads of R&D, the most relevant platform managers, the process owner of the product roadmap, a market specialist, and, if necessary, other specialists.

Technology Intelligence

The goal of TI at RIC is, in general, to make available relevant information about technological trends in the RIC environment in order to respond to opportunities and threats, and in particular, to provide this information to technology platform management and to technology roadmapping. To do so, RIC has adopted the Opportunity Landscape concept with its gatekeeper approach. How the Opportunity Landscape is used at RIC is the topic of the next section.

The Opportunity Landscape at RIC

Details of the Opportunity Landscape at RIC differ from the original concept at Straumann. However, the overall idea remains the same. In this section emphasis is given to the integration of the Opportunity Landscape in technology management, the roles of gatekeepers, and supporting tools. Then, implementation and practical experiences are illustrated briefly.

Integration of the Opportunity Landscape in technology management

The Opportunity Landscape covers TI at RIC. This is one of three technology management elements that have recently been implemented. There are direct interfaces to technology platform management and

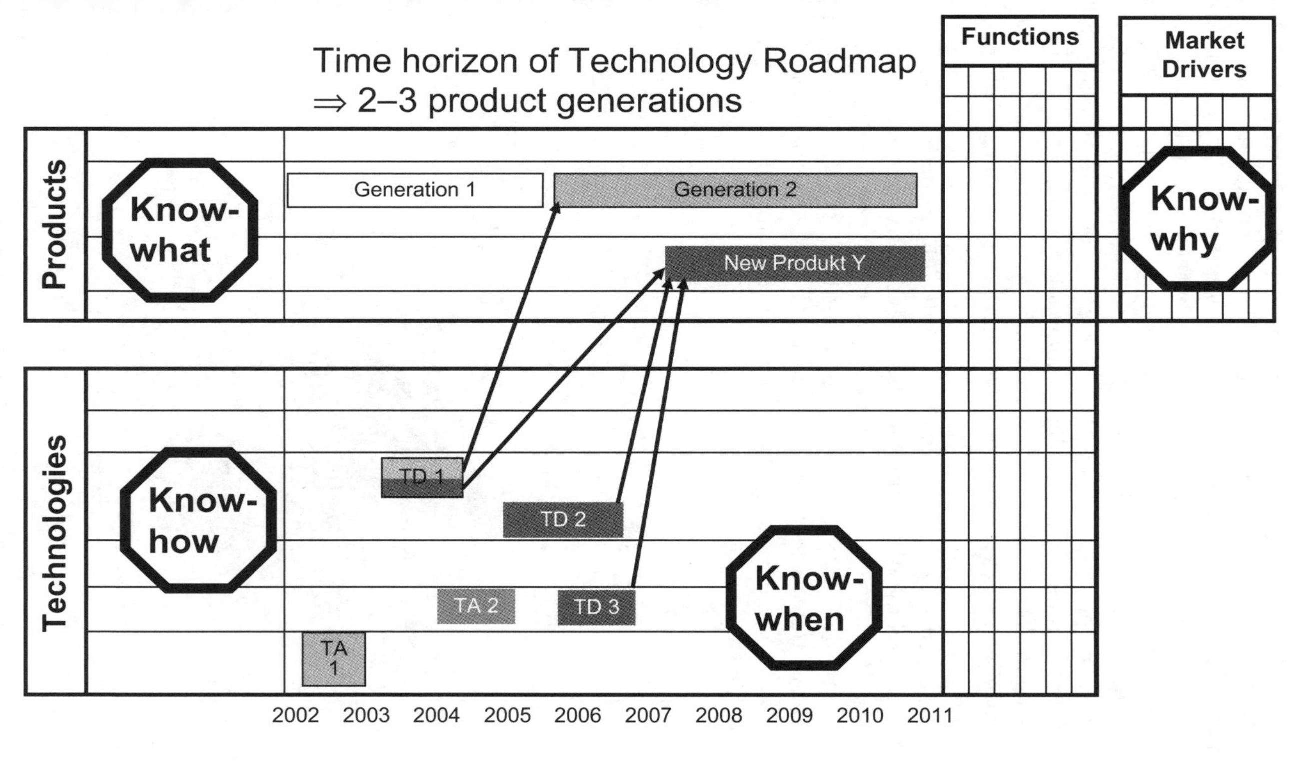

Figure 7.6 Technology roadmap at RIC, illustrative

technology roadmapping. A technology platform contains several knowledge and technology domains. These domains are issues in which RIC has high, moderate, or low competence (Figure 7.7). In order to maintain or expand competence in these issues, gatekeepers are assigned to them. Thirty-eight gatekeepers were defined at RIC. These gatekeepers are responsible for observing trends in their domains and providing information to the technology platform managers for input during technology roadmapping. In particular, during the second step of roadmapping, platform managers have to be informed by the gatekeepers about internal and external technological evolution.

Gatekeepers originate from RIC R&D units, RIC laboratory network solutions, and the RIC technology group. Emphasis is given to competencies; therefore gatekeepers in one technology platform may originate from different R&D units. This seems to be very promising for knowledge transfer between different units. All gatekeepers together build a gatekeeper network that is coordinated and animated by the head of technology.

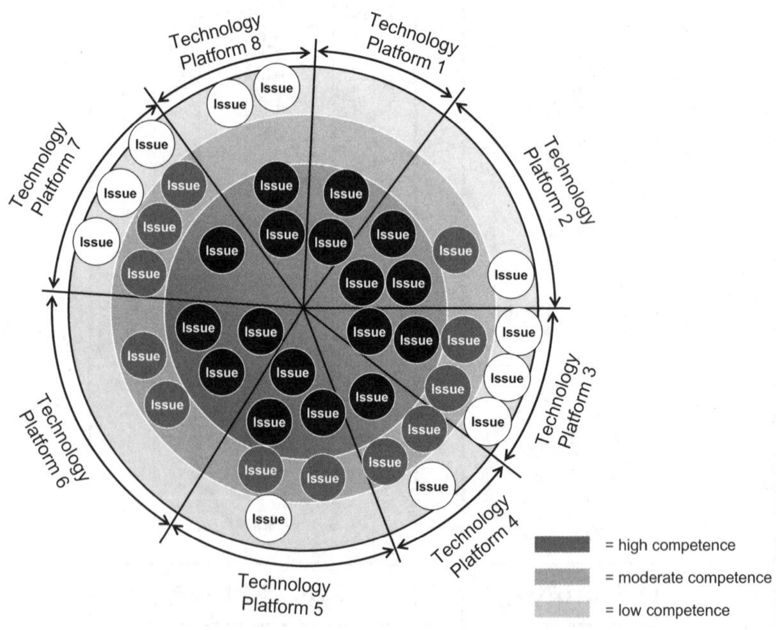

Figure 7.7 Visualization of the Opportunity Landscape at RIC, illustrative

Gatekeepers' tasks and communication

The tasks of a gatekeeper are to identify RIC-relevant facts and trends within the assigned issue, to assess these facts and trends, and to make insights available to other RIC employees. Accordingly, a gatekeeper is responsible for integrating knowledge from the company's environment, and giving expertise whenever needed. In order to identify relevant facts and trends, a gatekeeper has to continuously observe the technological environment by communicating formally and informally with experts, reviewing periodicals, attending seminars, etc. Assessment of these facts and trends follows a specific procedure and involves technology platform managers. Normally, gatekeepers and platform managers meet before the roadmapping process. However, if necessary, they may meet more often formally or informally in order to continuously be able to take initiative or to react to new insight.

Communication follows an information push and pull logic. Gatekeepers push information during various formal and informal meetings or discussions. Information pull is possible by directly contacting a gatekeeper for expertise. Since the list of gatekeepers is internally accessible (at least for R&D employees), experts who can help solve a problem can easily be identified. In order to be able to provide information, the gatekeepers should externalize their knowledge as often as possible, i.e. by filling templates.

Additional time exposure should be reduced to the minimum. Therefore, most gatekeepers are already experts within their issue. Some of them, in particular those from the RIC technology group, have already done gatekeeper tasks before. However, since some issues are very future-oriented, some gatekeepers have to spend a considerable amount of time on this task (at least initially).

Supporting tools and templates for gatekeepers

Gatekeepers are supported by several tools and templates. These templates assist gatekeepers along an idealized TI process. There are templates for information need formulation (information sources.doc), collection (insight/idea.doc), and analysis (attractiveness.doc, risk/cost.doc, competitive power.doc, and assessment.doc). Their interaction is illustrated in Figure 7.8.

From a gatekeepers point of view the two most important templates will be described in brief: information sources and insight/ideas:

The **information sources template** helps the gatekeeper to identify, structure, and manage her/his information sources (Figure 7.9). At the beginning, the gatekeeper should break down her/his issue by

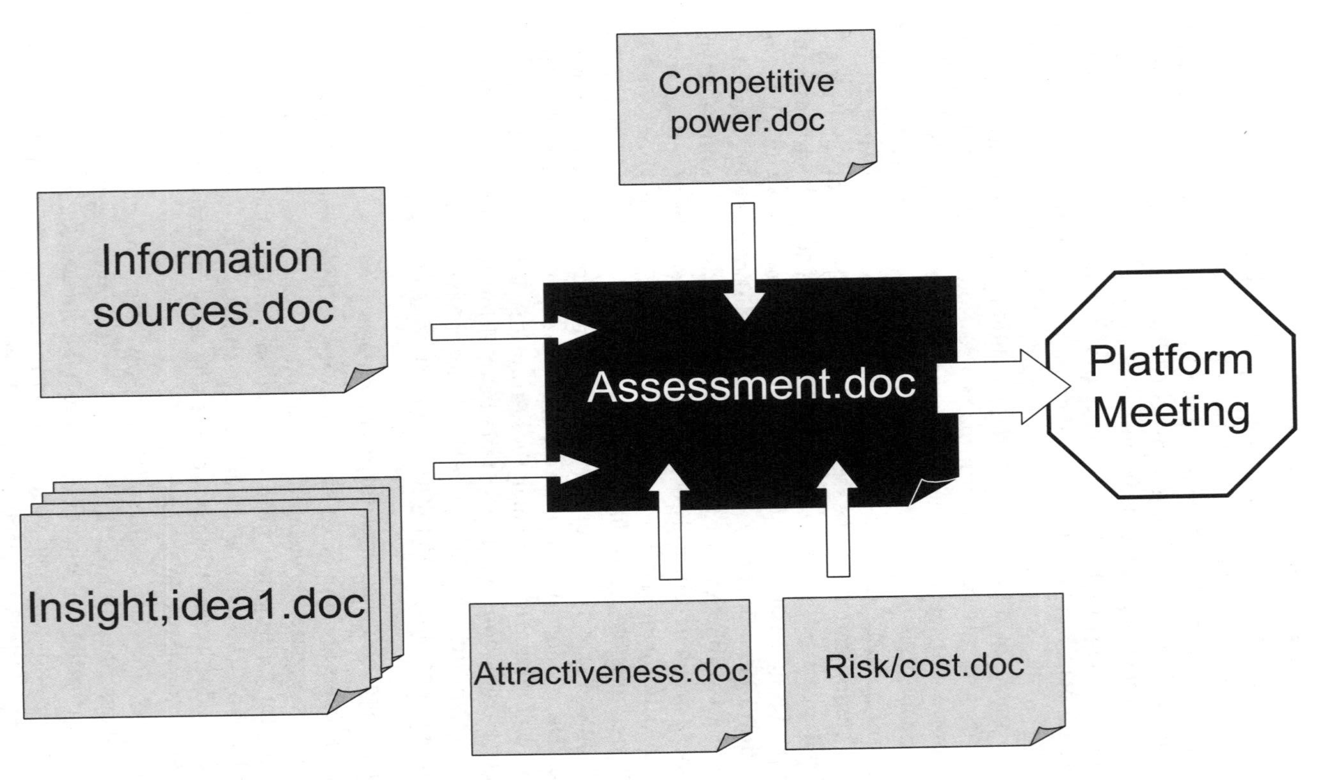

Figure 7.8 Technology assessment with gatekeeper information

determining subdomains that are of interest within the issue. The issue "optics," for example, may be subdivided into "spectroscopy," "interferometry," "integrated optics," etc. These subfields may be at different levels, i.e. they may cover different aspects of an issue. Then, the gatekeeper lists her/his personal formal and informal information sources as accurately as possible. For each information source, the number of consultations and the time needed for this consultation are estimated. This helps the gatekeeper to plan and control her/his gatekeeper activities. Finally, for each subfield and information source combination, the relevance and security is estimated. This helps the gatekeeper to identify gaps and to avoid redundancies. In addition, the quality label indicates the reliability of information and insight for further use, i.e. during technology roadmapping.

Next to the individual benefit, completing this template is important in case of gatekeeper change. Under those circumstances, existing information sources do not get lost and formal/informal networks can be maintained.

The **insight/ideas template** helps the gatekeeper to retain relevant information (Figure 7.10). In order to give the insight or idea a face, the gatekeeper defines a meaningful headline. Then, she/he writes down reflections following a proposed scheme, which is illustrative. The totality of all these notes should help the gatekeeper to be prepared when any information is needed.

All templates are understood as supporting tools. Complementing them is not mandatory but gives the gatekeeper an idea about in what kind of information RIC is interested. These templates are available online and off-line.

Complementary to the templates, gatekeepers have at their disposal a TI method toolbox. This toolbox is a description of various methods that can be useful for gatekeeper work. Again, use of these methods is voluntary. Typical methods in this toolbox are:

- Portfolios (for example, a technology position portfolio, a dynamic technology portfolio)
- Curves (for example, trend extrapolation, S-curves)
- Morphological box
- Relevance tree
- Frequency analysis (for example for patents and publications)
- Bibliometrics
- Scenario analysis
- Relation map

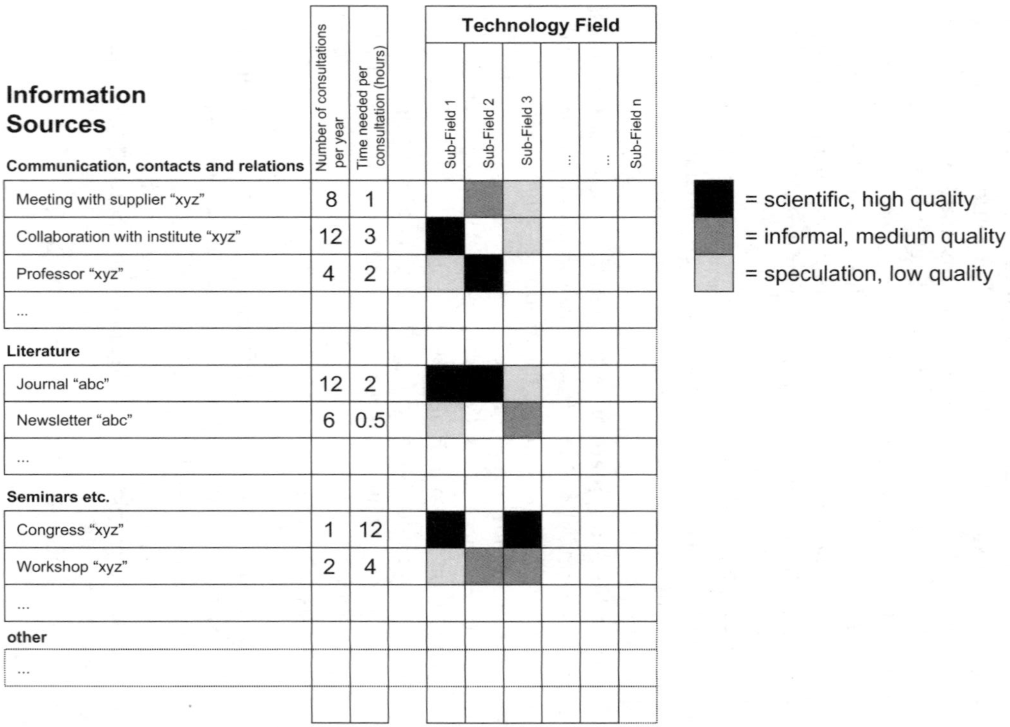

Information Sources	Number of consultations per year	Time needed per consultation (hours)	Technology Field: Sub-Field 1	Sub-Field 2	Sub-Field 3	...	...	Sub-Field n
Communication, contacts and relations								
Meeting with supplier “xyz”	8	1						
Collaboration with institute “xyz”	12	3						
Professor “xyz”	4	2						
...								
Literature								
Journal “abc”	12	2						
Newsletter “abc”	6	0.5						
...								
Seminars etc.								
Congress “xyz”	1	12						
Workshop “xyz”	2	4						
...								
other								
...								

= scientific, high quality
= informal, medium quality
= speculation, low quality

Figure 7.9 Gatekeeper template for information sources, illustrative

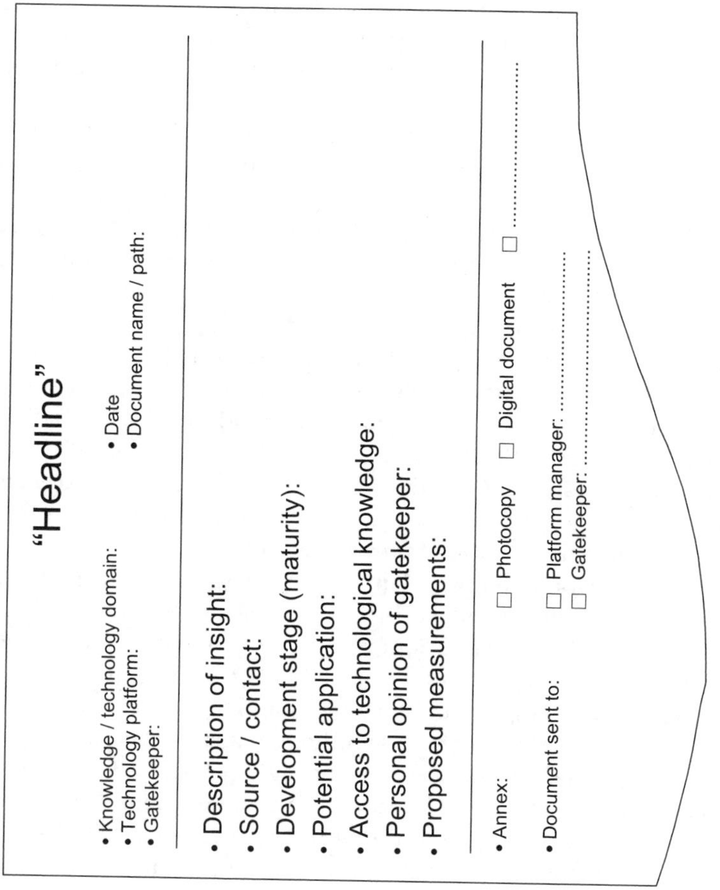

“Headline”

- Knowledge / technology domain:
- Technology platform:
- Gatekeeper:
- Date
- Document name / path:

- Description of insight:
- Source / contact:
- Development stage (maturity):
- Potential application:
- Access to technological knowledge:
- Personal opinion of gatekeeper:
- Proposed measurements:

- Annex: ☐ Photocopy ☐ Digital document ☐ ....
- Document sent to: ☐ Platform manager: ....
  ☐ Gatekeeper: ....

Figure 7.10 Gatekeeper template for insight and ideas, illustrative

Implementation of the Opportunity Landscape at RIC

Implementation of the Opportunity Landscape was straightforward. Because the concept has been adapted from the Opportunity Landscape at Straumann, development work was less time-intensive. Implementation was an integrated part of the technology management initiation project. Since all three elements – technology platform management, technology roadmapping, and TI – are interrelated, they were not considered as separate projects. Implementation of the Opportunity Landscape *per se* was the main topic during three workshops.

The **first workshop** was designed to define knowledge and technology domains (issues). Mainly the RIC R&D steering members and technology platform managers participated in this workshop. The previously determined technology platforms were the basis for issue definition. This is typically a top-down approach, because technology platforms had been derived from RIC technology policy and strategy (Figure 7.4). The issues are supposed to cover most important and crucial fields within a platform. Thus, by means of these issues, RIC maintains and establishes competencies in order to assure the long-term-oriented functionalities. In total, the workshop evoked about 40 issues. Basic technologies were, in principle, not considered in the Opportunity Landscape. Once the issues were determined, gatekeepers were tentatively identified. Hence, technology platform managers had to refine gatekeeper allocation for their platform.

The **second workshop** was planned to initiate and train the gatekeepers. In fact, the second workshop was held twice because of the large number of gatekeepers. Content and procedure were exactly the same. During these workshops the gatekeepers were first initiated into the overall technology management initiation project in order to understand their field of action, even though most of the gatekeepers had been informed by the platform manager beforehand. The gatekeepers' tasks were also explained and illustrated with numerous examples. Then, in the major part of the workshop, the gatekeepers had to deal with their issue for a first attempt. Initial tasks were to determine subdomains and to list existing and potential information sources. Most gatekeepers fulfilled these tasks in their platform group. Finally, two gatekeepers were selected to present their reflections in order to benchmark with others and in order to assure that all gatekeepers were speaking the same language.

The **third workshop** phase was to consolidate initial gatekeeper results. A workshop on each technology platform was organized. The goal was to prepare information for an initial technology roadmapping attempt. Therefore the gatekeepers had about one month to gain some insight about trends in their issues. Because of this short time span, information was not supposed to be exhaustive but to give the platform manager some first ideas about trends.

In summary, implementation of the Opportunity Landscape was a "piggy-back" project of the technology management initiation project. Since the concept of the Opportunity Landscape already existed, implementation was not a parallel-synchronous process to the design. However, the concept had to be adapted, which can be interpreted as development of the concept. Thus, implementation was parallel to development.

Practical experience with the Opportunity Landscape at RIC

Preliminary experience with the Opportunity Landscape at RIC is restricted to the workshops. Thus, practical experience originates from gatekeeper work during one month and may therefore not be representative. However, some observations could be made during this month.

The number of 38 gatekeepers seems to be high. The challenge was to **motivate all gatekeepers** during one single initiation workshop. There was a skepticism concerning additional time exposure. Some gatekeepers even argued that this activity was not agreed upon in the "management by objectives," and therefore refused to be gatekeepers. Some of these skeptical employees could be convinced that these activities are very favorable to them. One argument for this was that, since most of them already did these activities informally, perception of their work would be improved by the official gatekeeper status. A few gatekeepers who could not be convinced were replaced.

Giving the gatekeepers some time to **structure their issues** during the initiation workshop turned out to be very promising. Ambiguities could be cleared up, and after only one hour, most issues were structured and information sources were listed.

An interesting observation was made considering the **use of templates**. These templates, in particular the information source template, were considered very useful. The first applications during the workshops forced the gatekeepers to compare a targeted and actual state of information sources. Since during the workshop gatekeepers worked in teams, they could already identify redundant information sources, respectively they could add another gatekeeper as an additional information source for their own issue in order to profit from an interesting contact. RIC management wanted the templates to be more automated and directly linked (for example assessment templates). However, experience showed that over-automatization decreases flexibility and acceptance of tool usage.

Comparison of the Opportunity Landscapes at RIC and Straumann

As already mentioned, the basic idea of the Opportunity Landscape at Straumann and at RIC is the same, however differences exist in the details. Thus, implementation at RIC is insightful for improvement and for the flexible use of the Opportunity Landscape. The most important insight is interpreted below.

Content of the Opportunity Landscape

The nature of content of **issues** at RIC is comparable to the nature of content at Straumann. In both cases, issues are knowledge and technology fields which vary in level of aggregation (for example, nanotechnology and surface technology). At RIC, each issue was divided into subfields, which has not been done at Straumann. This division seems to be promising because the content of an issue, which is rather fuzzy at the level of its "headline," has to be structured in order to guide the gatekeeper in its activities.

The meaning of the "**slices**" of the Opportunity Landscape, however, differs considerably at RIC from those at Straumann. At Straumann, these slices represent strategic observation fields that just bundle the issues. These fields do not have a specific function, nor are they all at the same level (for example, materials, time-to-teeth etc.). At RIC, on the contrary, these slices represent technology platforms which, in turn, represent the main functionalities that RIC offers to clients. As opposed to the situation at Straumann, Opportunity Landscape slices are managed by the platform manager at RIC. The difference in the significance of the slices underlines the different approach of content definition. At RIC, this process was well structured and embedded in a superordinated technology management initiation project. Technology platforms were derived from RIC technology policy and strategy in a top-down approach. At Straumann, this top-down approach did not work because business mission and strategy did not allow for it to do so. Therefore, a bottom-up approach was chosen. The RIC approach seems to be favorable, because with this top-down approach, each issue is one fragment of a puzzle. At Straumann, issues were defined by brain-writing. Thus, the correlation between all issues is not obvious. The list of issues may even be incomplete.

Another difference is the **additional dimension** in the Opportunity Landscape. At Straumann, this dimension indicates the depth of observation, which corresponds to the interest or actuality of the issues ("players" = continuous and intensive observation, "substitutes" = regular observation, and "juniors" = being aware). At RIC, this dimension is used to indicate the level of competence in an issue (high, moderate and low). This allows RIC to manage their competencies to be maintained or to be built up in order to assure long-term-oriented functionalities. Thus, management of the Opportunity Landscape means management of competencies, and therefore becomes a part of the company's strategy process.

Gatekeepers' role

The gatekeepers' role is comparable in both companies. They are supposed to be experts and they have the task of observing facts and trends within their issue. However, the way they execute their tasks differs. At RIC, gatekeepers have to follow a well defined process, and they are supported with tools and templates. At Straumann, it is up to the gatekeepers to define how the task is best executed. There is just a sample annual report that indicates some questions that gatekeepers should be able to answer. Gatekeepers at Straumann can store information in a Lotus Notes database. Experience at RIC showed that templates were adequately used, whereas the sample report and the database at Straumann were barely used.

Gatekeepers' **communication** has common and different aspects in the two companies. In both cases, there is an information push and information pull logic. Pulling information means that any employee, in particular from R&D, may contact the gatekeeper for expertise. Information push at RIC brings information to the technology platform manager, at least twice a year for technology roadmapping. At Straumann, information is pushed by storing information in the database.

Coordination

Coordination of the Opportunity Landscape at Straumann and RIC differ considerably. At Straumann the Opportunity Landscape is very **democratic**, which means that it is the responsibility of all gatekeepers to assure that the Opportunity Landscape accomplishes its mission. There is a "central" coordinator, who animates the Opportunity Landscape. Animating in this context means motivating the gatekeepers, supplying them with some tools, and animating the IT platform.

The coordinator cannot exert pressure on the gatekeepers because he is not a "client" of the gatekeepers, nor is he hierarchically in a favorable position (he is a staff group member). Therefore, the Opportunity Landscape at Straumann is in a kind of lethargic state.

At RIC, the Opportunity Landscape is coordinated **top-down**. Already the definition of issues followed a top-down approach, and the first gatekeeper activities were guided by an external expert in this field. Gatekeepers had to deliver some results on exemplary workshops. There was a clear client during these workshops: the platform managers. Platform managers have to coordinate the gatekeepers within their platform in order to receive exactly the information she/he needs. If information is missing, it is up to her/him to take measures. Platform managers, in turn, are coordinated by the head of technology. He is permanently informed about activities of gatekeepers and, therefore, he supervises the Opportunity Landscape as a whole. He can avoid redundant activities and balance quality disparities.

Implementation

Implementation of the Opportunity Landscape was also different at both companies. At Straumann implementation was **explorative** and **synchronous** to the design of the Opportunity Landscape. RIC could profit from the existing concept, therefore implementation was less explorative, and also not parallel-synchronous to the design of the original concept. However, since the concept had to be adapted (which is equal to development of the concept) and contents such as issues, gatekeepers, information sources etc. had to be defined, one can speak about parallel development and implementation anyway.

This difference in implementation also explains the difference of the implementation project **duration**. While at Straumann this project took up to one year, at RIC the project only took a few months. This indicates that adopting and developing an existing solution makes sense. The shorter duration at RIC was also a result of a straightforward approach (which in turn was possible because a concept already existed).

In spite of the fact that in one company the original concept of the Opportunity Landscape was designed, and in the other company "just" developed, there were several parallels of implementation at these two companies. In both companies, implementation was done by means of several **workshops**. In particular gatekeepers were initiated in a group, not individually. The advantage of this is to have a common understanding and to save time. To give the gatekeepers some time during the workshop to reflect about their tasks turned out to be promising. Another parallel was that in both companies there were **champions**. In fact, the role and influence of the head of Technology Center at Straumann and the head of technology at RIC were comparable. They pushed and supervised the project. Also in both companies an external expert coached the project. He was the expert champion. Nevertheless, the "expert competencies" had to be shifted from external to internal. At Straumann, this competence was shifted to the TI coordinator, at RIC the head of technology and the platform managers were then responsible for championing the Opportunity Landscape.

Horizontal validation case B: screening process

Parallel to Straumann, Esec (Cham, Switzerland), in the context of a superior technology management project, designed and implemented a **technology and business opportunity process**, which is comparable to the screening process.

Numerous interviews and workshops were held during a time period of four months. All heads of business units, innovation managers, and project managers for large projects were involved during the first stage. Then, in the second stage, the focus was laid on two "exemplary" business units.

After a description of the company, the innovation management project will be illustrated in order to understand the context in which the screening process is embedded. The technology and business opportunity process at Esec and the screening process at Straumann will have some common and some differing characteristics. They will be compared and discussed. Again, the conclusions of this section will enter into the overall conclusion of the chapter.

Esec: company profile

The Esec group is a global provider of chip assembly equipment, processing techniques, and system solutions for the semiconductor industry. These machines and solutions are utilized in what is known as the back-end segment of semiconductor production. The core business of Esec encompasses Die Bonders (fastening a chip to a substrate material), Wire Bonders (establishing electrical connections between chip and substrate through the use of gold thread), Flip Chip Bonders (combination of die bonding and wire bonding), and Factory Integration (Figure 7.11). Geographic markets are Asia (62%), Europe (20%), and the USA (18%).

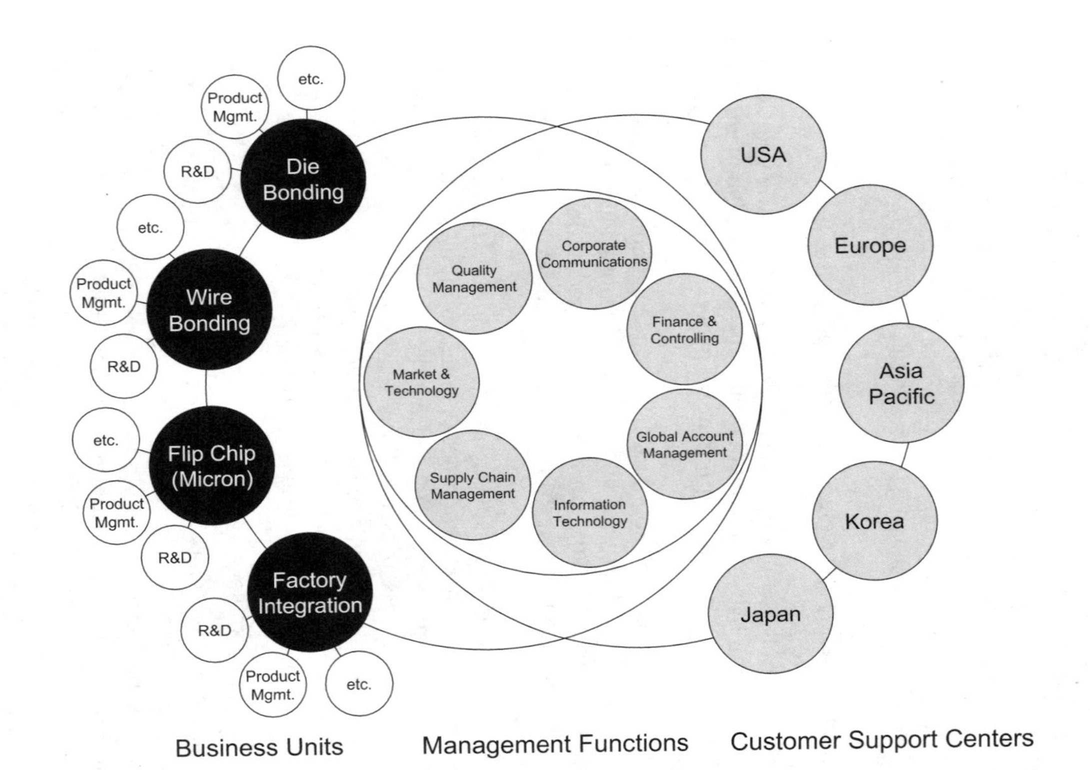

Figure 7.11 Structure of the Esec group

Due to the very cyclical semiconductor market (for example the collapse of prices in the semiconductor industry in 1998, and worldwide overcapacity), net sales and head count vary considerably. The latest net sales in 2000/2001 were about US$650 million, head count in 2000/2001 was 950 employees.

The ESEC group's business mission is:

> *We provide equipment and system solutions that are unique in the world when it comes to the assembly of semiconductors and their placement on printed circuit boards.*

Esec is committed to being a technological leader in the field of equipment manufacturing in the back end of the semiconductor industry by "setting standards."

Innovation management at Esec

R&D at Esec is mainly project-oriented. In fact, the focus of R&D is on development and engineering, which is performed in business units (Figure 7.11). However, a process model for product innovation is valid for all R&D projects. The process model envisions diverse processes that are run synchronous and asynchronous-parallel, and sequentially (Figure 7.12).

The technology and business opportunity processes are comparable to the screening process at Straumann. The presentation of these two processes is the content of the next section. In this section, some other important processes that are of particular interest to the technology and business opportunity processes will be briefly described. These additional processes are the strategic planning process, the project management process, and the product development process.

Strategic planning process

The strategic planning process involves interaction between the technology and the business opportunity process. On the one hand, the strategic planning process defines the field of action of the opportunity processes, and therefore delivers the basis for decision-making for product innovation. On the other hand, the technology and business opportunity processes deliver new insight for the strategic planning process, in particular for technology strategy.

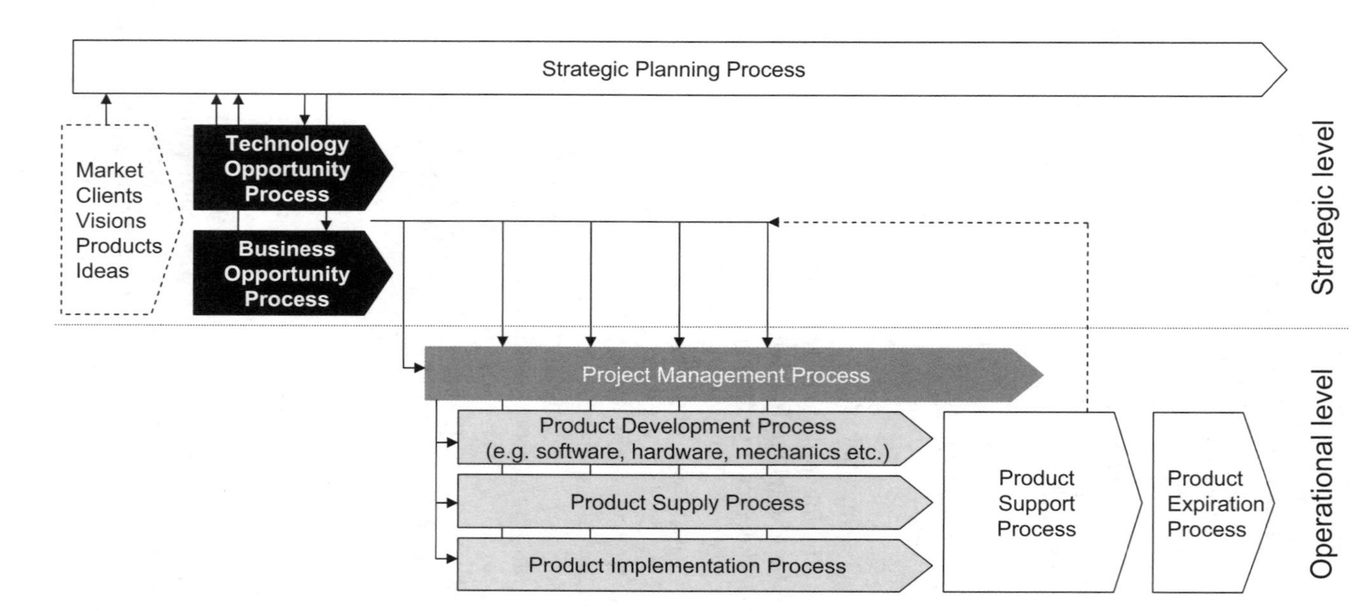

Figure 7.12 Process model of product innovation

Project management process

The project management process describes all tasks that the project leader has to fulfill during product development. These tasks are depicted in Figure 7.13.

After project structure and process planning, the development project in the narrower sense can be released by means of a kick-off meeting. All information (and intelligence) that has been generated before is transferred to the concerned project members. Other tasks of the project leader are controlling and reporting. The understanding of controlling at Esec is a periodic comparison between the target and actual state of the project. Targets were set during project planning. Finally, the pilot project is coached. The pilot project provides the opportunity to test a new product the first time in a representative environment.

The project management process, of course, is owned by a business unit. A constant interaction with the market and technology group, which is the owner of the technology and business opportunity process, guarantees the link to new technologies and market changes.

Product development process

The goal of the product development process is to realize the targets that were set in the requirement specification. To achieve this goal, the project team has to work with the allocated resources, and to consider targeted project costs, deadlines, and quality requirements.

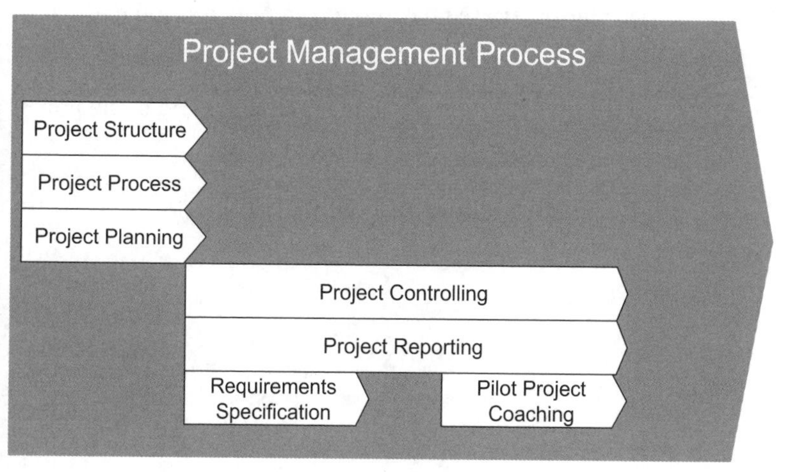

Figure 7.13 Project management process at Esec

The product development process consists of four main phases: the concept phase, the functional model phase, the prototype phase, and the initial batch phase (Figure 7.14).

During the concept phase, preliminary ideas for solutions and initial experiments are explored. Parallel to these initial steps in product development, the requirement specifications are set in the project management process. Thus, there is an intensive interaction. Also in parallel to the concept phase, the feasibility is explored in detail during the functional model phase. Subsequently, a prototype is designed and produced. This phase is divided into two phases. A first evaluation is done internally (alpha test), a second evaluation is done in collaboration with a "lead user" (beta test). Knowledge from these tests allows the production of an initial batch in order to test the product on a broader basis. The transfer from the initial batch to mass production terminates the product development process.

It would be favorable to link all product development processes at Esec, in order to profit from synergies, and to avoid redundancies. However, interaction between different product development processes seems to cause problems at Esec because project natures differ considerably.

Screening (technology and business opportunity) process at Esec

The technology and business opportunity processes at Esec have the same goal as the screening process at Straumann, which is "to lead to effective and efficient decision-making before product development." However, the solution at Esec has some common and some different characteristics compared to the solution at Straumann. This section first describes the stages of the technology and business opportunity processes and is then followed by a brief explanation and discussion of

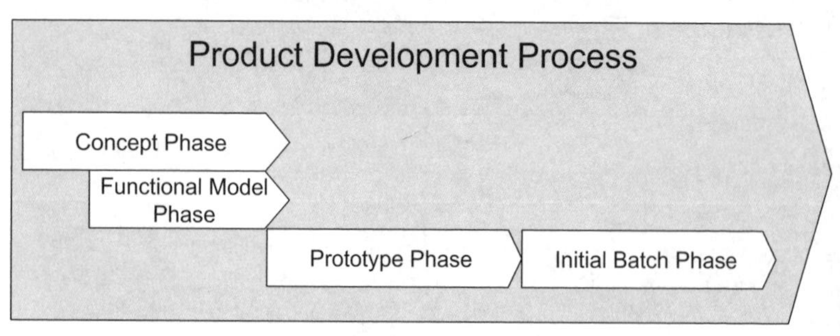

Figure 7.14 Product development process at Esec

the coordination of the processes. Finally, a short description of the implementation of technology and business opportunity processes is given. No insight into practical experience could be gained at Esec. Thus, the content of this section is a reflection of reality in practice, but is not thoroughly tested in practice.

Stages of the technology and business opportunity processes

The technology and business opportunity processes take place in the early stages of product development. These two processes are idealistically divided in order to emphasize an equivalent technology and market view. In practice, the two processes are interchangeable and a strict division cannot be observed. The input into the processes emerges from competitors, markets, customers and other sources from the company's environment, as well as from strategy, internal ideas etc. The output is a business plan with requirement specifications and other basics for effective and efficient project development. The opportunity processes present three stages and gates: The first stage is to develop ideas. The second stage contains market and technological (feasibility) research. Finally, the third stage defines the product. Between the stages, gates "officially" synchronize the two processes, and the opportunity passes into the subsequent stage (Figure 7.15).

The most important elements in theses stages and gates, and which are of particular interest with regard to TI, will be described in the following. The emphasis is on the technology opportunity process.

Stage 1: Develop ideas

The stage "develop ideas" is very close to the classic intelligence cycle, which contains interactive and iterative phases. The **definition of innovation fields** fits with defining the observation area and intelligence needs. This is combined with business mission and strategy. To **identify and collect technology (market) data** is the information collection stage. Esec differentiates between an undirected search (scanning) to identify new technological facts and trends, a clearly defined search (monitoring) to observe yet identified technologies, and the sourcing of technologies (scouting). To **analyze technologies (market data)** structures the information. The most important output of this phase is the status of the technological development (S-curve). During **assessment of technologies (market data)** the previously analyzed technologies are put into the Esec context and positioned into portfolios. Then, to update and inform technology corresponds to the information dissemination phase. This build the basis for decision-making.

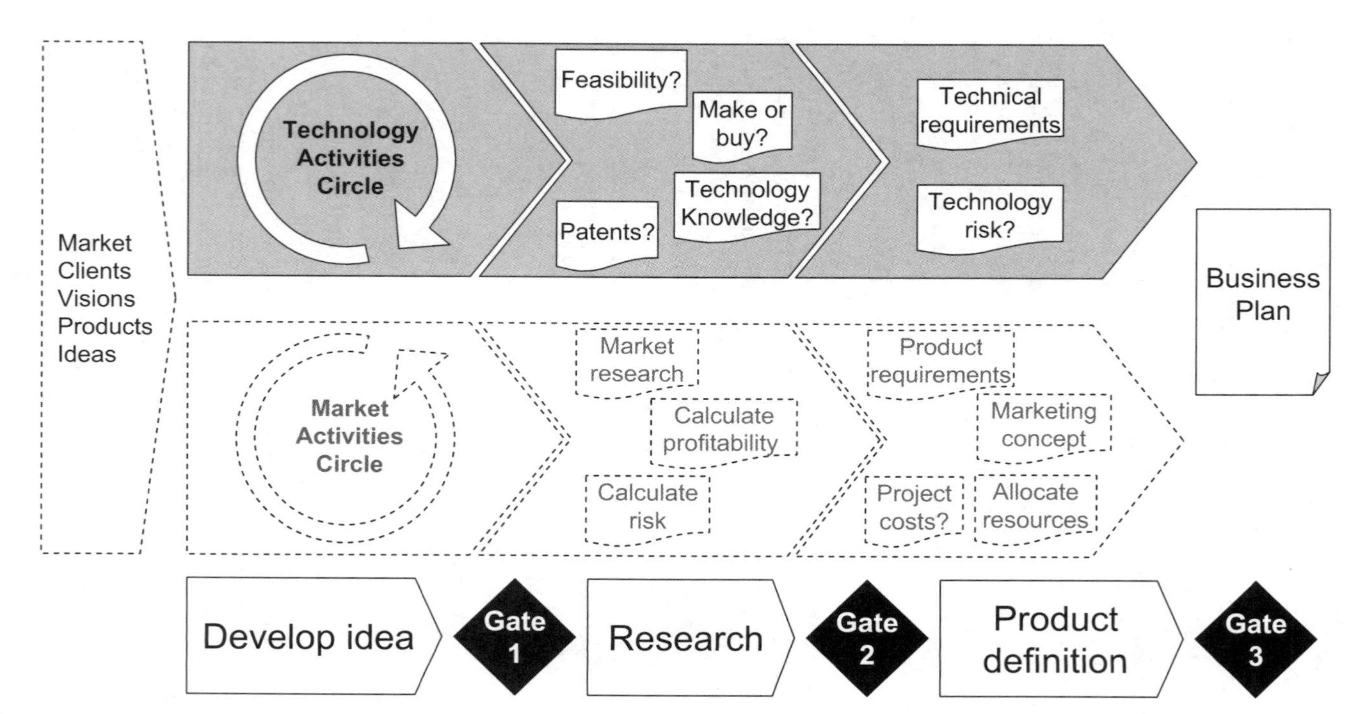

Figure 7.15 Stages and gates of the technology and market opportunity processes

Finally, merge and create ideas brings together technology and market information (handshake), which builds the basis for generating a new product idea.

Gate 1: Release research

The approval for research to proceed is a first synchronization point between the technology and the market view. The goal is to formulate and approve a **business opportunity description** which presents the potential for a successful product. There is a detailed check-list for this release. If the decision is positive, the idea passes to the research stage for further investigation.

Stage 2: Research

During the research stage a **feasibility study** is initiated. In fact, this study continues up to the end of Stage 3. The goal is to build the basis for the technical requirements for the business opportunity. This feasibility study should show in particular, if the technology can be developed in-house. Furthermore a first (lean) prototype should be developed. In addition to the feasibility study the **patent situation** must be cleared. Both possible patent application and patent infringement are to be checked in order to protect intellectual property, and to avoid both problems and time delays. Then, also considering the check of the patent situation, a **make-or-buy decision** is made whether the technology should be developed in-house (make), be purchased externally (buy) or a combination of both. Basically Esec is open to all forms of cooperation. Finally **technology knowledge is built-up**. This is obviously a consequence of the make-or-buy decision. The goal is to prepare technology knowledge for product development in a narrower sense. Therefore knowledge is transferred from the market & technology group to business units. From a market view, **market research** is done, and **profitability** as well as **risk** are calculated during this research stage.

Gate 2: Release product definition

The objective of this gate is to formulate and approve the **product definition**. All basics were studied during research. Again, if there is a positive decision, at the last stage, product definition can be initiated.

Stage 3: Product definition

During this product definition stage the **feasibility study** is continued and finalized. The goal is to build knowledge to the extent that during the subsequent product development process, risks would be minimized and time delays would be avoided. Again, the goal of the finalized feasibility study is to present a prototype. Increasing requirements of costs, time-to-market and quality increase the risks. Therefore the **technology risks** are identified, assessed and, if possible, minimized. Next to the definition of the **product requirements** from a business point of view, the creation of a **marketing concept** and the calculation of **project cost**, **resource allocation** is another element during this product definition stage.

Gate 3: Release project

Gate 3 concludes the product definition and **releases a project**. A business opportunity committee decides on the realization of the product, which means a transfer of accumulated knowledge during the technology and business opportunity process to the product development process. The most important element for this transfer is a business plan.

Coordination of the technology and business opportunity processes

The principal process owner is the "market & technology manager" (cf. Figure 7.11). She/he coordinates and coaches tasks in the processes and organizes workshops. The processes are performed periodically, about four times a year at Esec. In addition to the central market & technology manager, each business unit has its market & technology group. These groups are not fulltime jobs but tasks done by some employees in addition to their daily responsibilities. This organization makes the technology and business opportunity processes a support for business units, which is accepted by means of participation. In addition, knowledge transfer does not present a major concern, because most knowledge is developed in the business units. Because the processes are coordinated centrally, redundancies can be avoided, and synergies promoted.

Discussion of the technology and business opportunity processes

The organization of the technology and business opportunity processes is well defined. Such an organization allows all inputs to be judged equally. Very interesting seems to be the differentiation between assessment of ideas and assessment of projects. An evaluation of ideas only after their definition as a project would limit the openness to new ideas and would delay a refinement of the ideas. The separate management of ideas and projects, in the form of separate portfolios and a corresponding roadmapping could bring clarity into the evaluation process. However, such a well structured process may also restrain some ideas because of its "official" and irregular character. The official character may be hindering because there is no room for "crazy" ideas. And the periodic process, and by this the infrequent ideas assessment – at Esec about four times a year – may demotivate people who have posted an idea.

The previously described processes are very similar to what literature refers to as the "fuzzy front-end" of innovations (Figure 2.11). The main difference between the Esec process and the process depicted in Figure 2.11 is that the former envisions Technology Intelligence activities exclusively during the "creating idea" stage, while the latter considers TI a continuous and parallel process during all stages. This is a fundamental difference, because at Esec, TI becomes project-oriented and provides the initial step for new product development. Therefore structures and responsibilities are only defined with this in mind, although a permanent and systematized Technology Intelligence seems to be promising.

Implementation of the technology and business opportunity processes

The technology and business opportunity processes have been partially implemented at Esec in a two stage proceeding. Since Esec did not have a technology management before, one goal was to sensitize people to the benefit of technology management. The focus was on the very first stages of the processes, i.e. the creating ideas stage. At the same time some tools that had been identified and evaluated before could be tested. Again, implementation was conducted by means of workshops.

The **first workshop** was conducted with the "flip chip" business unit Micron. This business unit was chosen because there was a need to have an overview of technologies, and because this business unit should become a new orientation. The fact that the technology and business opportunity processes could be tested "online" was a favorable circumstance. The workshop took five hours and, in addition to the workshop coaches, participation was limited to three people: the head of Micron R&D, a product manager and a software developer. The main result was the establishment of two technology portfolios (Figure 7.16).

The **second workshop** was conducted with the business unit "wire bonder" one week later. The content was similar to the first workshop. However, experience from the first workshop improved the quality of the second workshop. The head of the business unit, a wire bonder process engineer, two product manager and a project manager participated. Again, two portfolios could be established (Figure 7.16).

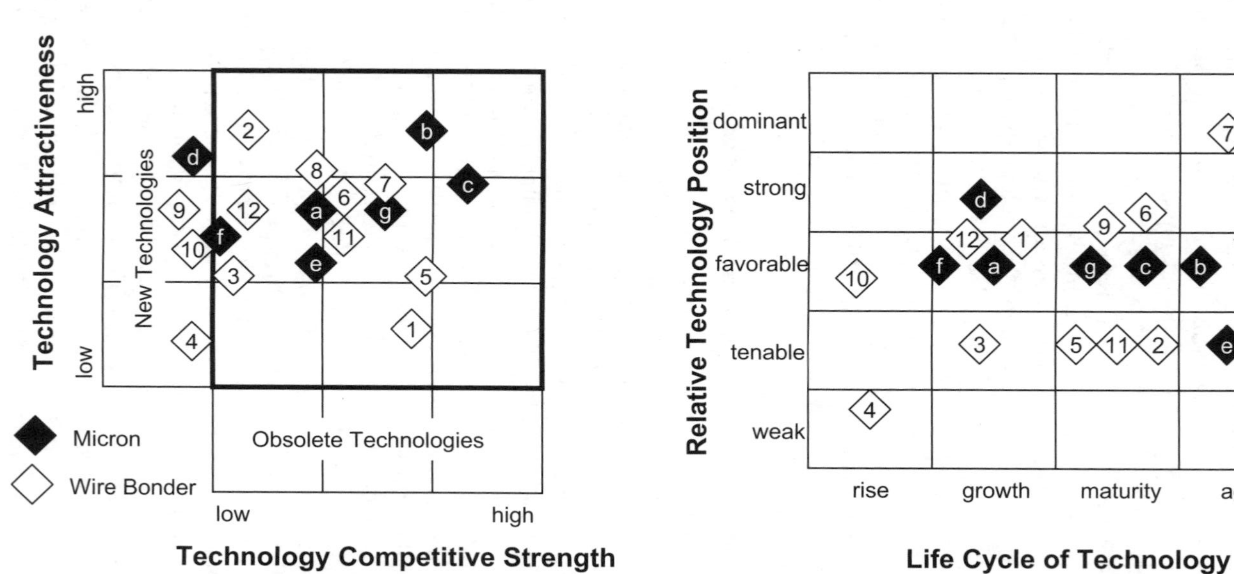

Figure 7.16 Technology portfolios as a result of workshops at Esec

Insight from implementation of the screening process at Esec is restricted to initial experiences. There is no valuable practical experience that could give insight about effectiveness and efficiency of the technology and business opportunity processes. However, the major insight is that implementation was strictly separated from design. Thus, this very well structured process has to be tested and adapted (i.e. "developed") with pilot projects. Another important insight is that workshop participants are open to this technology management approach. This lets us conclude that the situation for further implementation is favorable.

Comparison of the screening processes at Esec and Straumann

As mentioned before, the "screening process" at Esec has some common and some different characteristics compared to the one at Straumann. A discussion of these characteristics allows insight to be gained for improvement of a TI process that is very close to product development. The most important insight is interpreted below.

Content and organization of the screening process

The **goal** of the screening process at Straumann and Esec is to build the basis for product development. Input, which are ideas, proposals, etc., and output, which is a business plan or a project proposal, are the same at both companies. However, the process between input and output differs.

First of all, the process at Straumann consists of two **stages** which differ in terms of depth of examination. The goal of the first stage is to select, and thus to reduce the number of inputs. This is quite a reactive process. The goal of the second stage is to investigate in detail whether the input is worthy of being transformed into a project or not. At the end of this stage, the idea is still "on paper," in particular first reflections about technical solutions have not yet been developed. In turn, at Esec the process consists of three stages, which differ in terms of content. The goal of the first stage is to create an idea, which is more proactive than at Straumann. The goal of the second stage is to prepare product definition. This stage can be compared to the investigation stage at Straumann because technical as well as market aspects are explored. Finally, the goal of the third stage at Esec is to build technical knowledge, and eventually to develop a prototype. This content is part of the product development process at Straumann rather than done in the screening process. The stages and criteria at Esec are described in detail. An advantage of this is that the process is very complete, which guarantees that all inputs are treated equally. The disadvantage of such a detailed process is occasionally an unnecessary waste of resources, and the fact that reaction to unforeseen situations may cause difficulties. However, this detailed description is in line with an overall process-orientation in innovation at Esec. As opposed to Straumann where innovation is driven by the company culture. Therefore, the screening process is as detailed as necessary. The focus is on interdisciplinarity rather than on strict criteria.

An important difference between screening at Straumann and screening at Esec is **continuity** of the processes. At Straumann, screening is a permanent task. The advantage of this is that screening is a part of daily business, and inputs are examined without a delay. At Esec, screening is run about four times yearly. This can be explained by the stage-gate character of the process itself and the overall innovation process. Another explanation would be the fact that Straumann reacts on inputs in screening, Esec takes the initiative proactively in order to seek opportunities. At both companies the screening processes are intelligence processes. The nature of the differences in the processes at Straumann and Esec can be illustrated with Lichtenthaler's (2000: 347) view of the TI process. While at Esec the screening process consists of a **conveyance** and a **convergence** stage, at Straumann the process can be classed only as converging. Conveyance at Straumann is the task of the Opportunity Landscape. Thus, from a content point of view (not fundamentally) the process at Esec can be interpreted as a combination of the Opportunity Landscape and the screening process at Straumann.

Finally, an obvious difference is the **separation** of the screening process into a technology opportunity process and a business opportunity process at Esec. The reason for this is to give both views an equal weight. This differentiation corresponds to the process-orientation in innovation at Esec. At Straumann, a separation of the technology and the business view does not seem to be necessary, because company culture allows ideas to always be handled in a holistic context, which means that technology, market, strategic and organizational aspects are treated equally. Insight about common and different characteristics of the content of the screening processes at Esec and Straumann are illustrated in Figure 7.17.

Roles in the screening process

A fundamental difference is that Esec is organized by business units, Straumann just functionally. This also changes participation in the screening process. However, this difference is not so important because

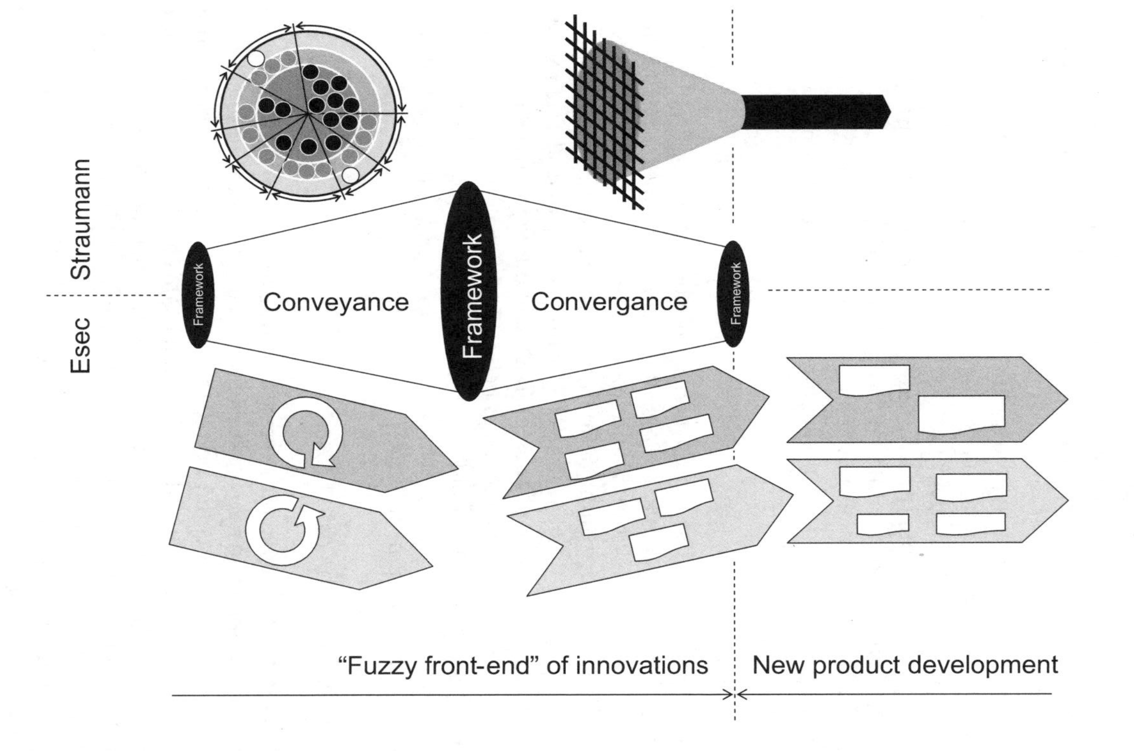

Figure 7.17 Common and different characteristics of the screening processes at Esec and Straumann

at Esec within the business units the multidisciplinarity view is also guaranteed through participation of people from different functions (product managers, engineers, etc.).

Therefore, roles at both companies are comparable: a central coordinator and animator, and a decentralized screening worker. One difference is that the coordinator at Straumann is a member of the technology management group, which is embedded in the Technology Center (R&D and product marketing), while at Esec the coordinator is in the market & technology group, which reports directly to the CEO. This gives output from the screening process at Esec more weight than at Straumann. Another difference is that Straumann has a permanent screening core team that completes competencies for further investigation. At Esec, screening teams are formed ad hoc due to the infrequent nature of the process.

Implementation

Implementation at both companies cannot be compared directly, since at Esec the processes had only been partially implemented. However, parallels are that in both cases implementation was done by means of workshops and that implementation of the screening process was a part of a superordinated project (Technology Center at Straumann, technology management at Esec). The main difference was that at Esec the process was designed in detail before implementation, while design and implementation could not be strictly separated at Straumann. However, during implementation at Esec there was still room for adaptation and development of the process.

Vertical validation case C: start-ups

The difference between this vertical validation case and the previous two horizontal validation cases is that insight is restricted to observation, i.e. no action could be taken. Therefore, the statement of this section is also different. In fact, the aim of this section is to discuss the existence and practice of TI in small companies, and how the generated elements of the solution at Straumann could be valuable for very small companies, i.e. start-ups.

A sample of 13 start-up companies was studied. In fact, this small sample does not allow statistical significance to be gained, but does provide an overview. Information in this section originates from in-depth interviews with the founders and directors of the examined start-ups. This gave the first insight into how these companies cope with this topic. Thus, after a short description of the sample, some trends in the practice of TI in start-ups will be described. Then, one solution (Zeptosens) will be presented in order to give an exemplary model of how a start-up can organize TI. Finally, the solution at Straumann will be compared to insight from the start-up study in order to discuss the validity of the generated TI elements with regard to company size.

Sample of start-ups

The interviewed sample consists of 13 start-up companies, mainly in the BioTech (6), MedTech (4), and IT/Electronics industry (3). A description of the sample start-ups is given in Table 7.1. Next to the year of foundation and actual number of employees, the field of action is briefly described by the company's vision (if existing) or their main products. Turn-over and R&D intensity are not listed because they are not comparable; some companies do not yet have any turn-over, some companies spend all employed resources on R&D. However, all companies are "technology-based" as defined in Chapter 2. It is also common to all these start-ups that they are explicitly (for example by the vision) or implicitly (for example by action) committed to being technological leaders.

Almost all BioTech companies were founded within the last five years and have more than fifteen employees. This indicates the enormous growth-rate in this industry. Typically, these start-ups are "spin-offs" from universities (ETH or the University of Zurich), or from large pharmaceutical companies (Novartis or Roche). Therefore they benefit from very specialized knowledge in a specific technology field. In addition to this some spin-off founders have years of experience within their field of activity. The field of activity is typically in a niche market and the company often has a restricted or even single product focus. Revenues (if any) of these start-ups typically originate from royalties from licensing out technologies or products. Selling market products seems to be possible when the start-up passes the 10-year mark.

These characteristics are comparable for MedTech start-ups. A difference is in the number of employees. Most examined MedTech start-ups have less than 15 employees. This might be explained by the fact that MedTech start-ups tend to be financed by a "weaker" partner than BioTech start-ups, which often benefit from large pharmaceutical companies' VC funds. Finally, IT/Electronics start-ups vary considerably. Most of them are more than five years old, number of employees vary between a few to 25. They are typically ETH or university spin-offs.

Table 7.1 Sample of start-ups

Start-up	*Industry*	*Foundation*	*Employees*	*Field of action, vision*
Arpida	BioTech	1997	34	"Research and development of anti-infectives"
Artificial Sensing Instruments (ASI)		1989	2	"Optical biosensors for diagnostics and environmental analysis"
ESBATech		1998	17	"Enabling faster drug discovery through cellular screening"
Genedata		1997	75	"Bringing data to life with the leader of bioinformatics"
The Genetics Company (TGC)		1998	27	"Innovate and accelerate the path from drug discovery to the clinic"
Zeptosens		1999	28	"Bioanalytic solutions"
Degradable Solutions) (DS	MedTech	1999	8	"Development of innovative medical therapies and substitution of permanent implants through the use of degradable solutions"
Medizintechnik Basel (MTB)		1989	10	"Ultrasonic diagnostic products"
ndd Medizintechnik		1992	11	"Spirometry in the digital age: pulmonary function testing"
Opthalmics Development Company (odc)		1997	7	"Diagnostics and therapeutic instruments and equipment for opthalmology and eye care"
NetVision	IT/ Electronics	1996	7	"From vision to reality: internet applications"
PI Electronics (PIE)		1995	8	"Industrial process instrumentation"
Schmid & Partner Engineering		1994	25	"Electromagnetic near-field scanning"

Practice of Technology Intelligence in start-ups

In order to gain insight into the practice of Technology Intelligence in these start-ups, one open question was formulated: "How do you get relevant information from the technological environment into your company?" The motivation behind the choice of this open and general question was to make the interviewee tell a "story," i.e. to get an unbiased answer. Whenever useful, additional questions were asked in order to gain a deeper understanding.

During this research three basic practices of coping with "getting external information" could be observed. One "group" of start-ups is aware of executing TI activities and they organize themselves, which is a systematic approach. Another group in reality undertakes TI activities, but is not aware of doing so, i.e. there is no system of organized TI activities. A third group has no TI activities at all. This basic differentiation explains **how companies cope with Technology Intelligence** (not how they are organized).

Furthermore, the question about **who pursues TI activities** gave the insight that either responsibility for TI activities is delegated to one single person, who is typically the CEO, or it is up to more than one or even all employees to fulfill this task. The first case has a centralized character, the latter is decentralized.

The distinction between these characteristics led to the definition of five models of how start-ups perform TI. These models will be separately discussed in short. A classification of the interviewed start-ups is illustrated in Figure 7.18 (dark grey: BioTech, medium grey: MedTech, white: IT/Electronics, in brackets = number of employees). Some "clusters" can be observed. However, since the sample is restricted and since none of the industries present a clear homogeneity, these clusters do not seem to be relevant. Another observation is that obviously bigger start-ups tend to share TI activities among a certain number of employees, whereas in smaller start-ups TI seems to be at the CEO's responsibility.

The **first type** is companies that are particularly interested in getting information from the technological environment, and therefore have a systematized approach. In this type, one person is in charge of this task. Only one company is classed in this type: ndd. The person in charge is the CEO, who is the head of R&D at the same time. The goal of his observations is to "keep abreast" with what happens in this industry. To do so, he specifically asks experts in this field about trends. However, the actual business focus is to bring products to

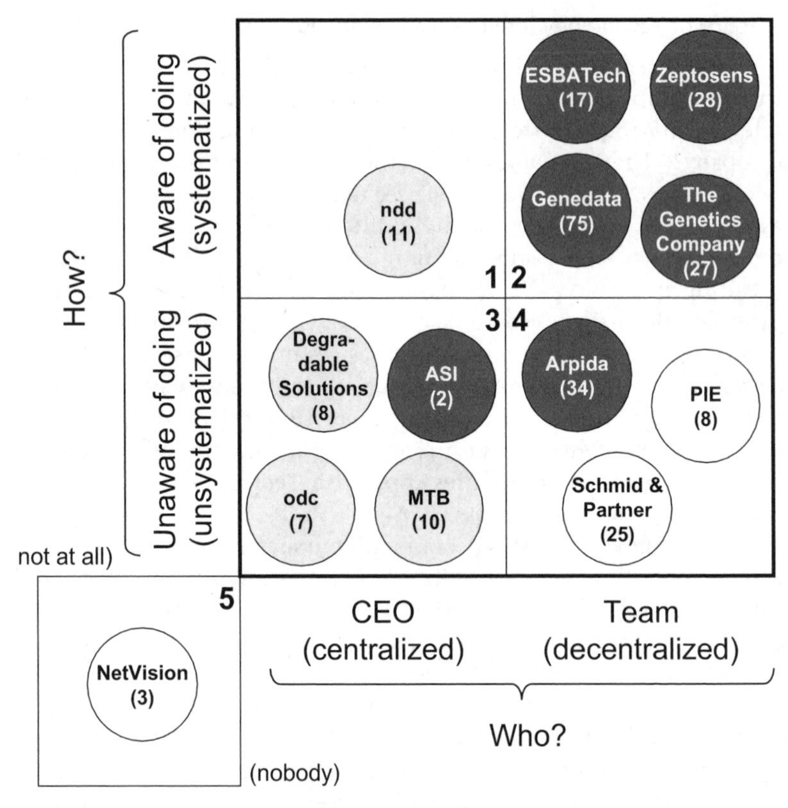

Figure 7.18 Organization of Technology Intelligence in start-ups

market (Europe and USA). Therefore, new product development and thus, trend information is not a major concern.

The **second type** is the most "sophisticated" type in start-ups. This type represents a systematic and a team-oriented Technology Intelligence. Four start-ups can be attached to this type: ESBATech, Genedata, TGC and Zeptosens. They are all active in the BioTech industry. At ESBATech senior scientists meet once a week. Each of them has a technology field in which she/he has to be informed about trends and to seek specific technologies. One agenda item is to discuss these trends. In addition to this, potential competitors (small and large companies) are systematically tracked. At Genedata there is a news group named "newswatch" on their intranet. The goal is that any employee can post a note whenever she/he makes an observation that might be relevant for Genedata. The nature of the observation can be scientific, political or market focused. Some employees are obliged to regularly post comments about their topic. At TGC, the executive board consists of four scientists (three of them former university professors). They estimate their situation as being a unique technological leader in their field of action. Their advanced knowledge and the permanent contact to their scientific advisory board guarantee that they will be informed about their technological environment. However, they consciously observe what other companies and universities do in their field of activity. Finally, Zeptosens is probably the most organized in TI. Because they present an excellent example of how a start-up can be organized in TI and because this second type is the most sophisticated, the TI system at Zeptosens will be exemplified in a separate section.

The **third type** is start-ups in which one person, mostly the CEO, does some TI activities, but is not aware of doing so. Four companies matched this description: ASI, DS, MTB, and odc. In fact, the CEOs, who are scientists, inform themselves by reading journals and by attending trade fairs. The aim of the latter is to sell products, thus TI is just a side-effect. However, the result is that the CEO is informed about trends in the technological environment.

The **fourth type** is companies in which all employees or a group (i.e. R&D members) haphazardly inform themselves about the state of the art in their field of action. Typically IT/Electronic start-ups belong to this type: PIE and Schmid & Partner. In addition to them, Arpida fits with them. Common to all three companies is that they are very product oriented. Therefore the focus is not on getting information for new product development, and thus, a systematic observation has only a secondary priority.

Finally, the **fifth type** is start-ups that have no activities at all with regard to Technology Intelligence. Only one company can be attached to this type: NetVision. In fact, this company estimates that it is not crucial to do TI activities in its field of activity, which is IT/Electronics.

Example of Technology Intelligence in a start-up (Zeptosens)

The brief description of TI practice in some start-ups showed that there does not seem to be a common "practice in use" of Technology Intelligence. Thus, normative guidelines do not seem to make sense, nor would it be possible because of the restricted sample. In turn, one example that seems to be a promising approach for a start-up offers some ideas.

Company description

Start-up: Zeptosens started its operations in March 1999. The multidisciplinary Zeptosens team – with a scientific background in physics, chemistry and biology – provides the skills and know-how to efficiently develop analytical nanotechnologies and methods as complete solutions. Today, Zeptosens employs 28 persons. Zeptosens is located in the Technology Center Witterswil, in close proximity to Basel, in the trinational Swiss/German/French region.

Focus of Activities: Zeptosens envisions initiating a new era in highly multiplexed, automated and ultrasensitive biomolecular analysis on nanotechnology-based microarrays and readout systems. Zeptosens is developing and will introduce highest sensitivity detection technologies for the analysis of a few hundred molecules on a microarray. With the Zepto™-Technology, the industrial user will obtain a tool to measure nucleic acids at highest sensitivities or to determine families of proteins all at once in order to gain rapid insight into biological processes and use this information for efficient product development.

Product Line: With the ZeptoTM-product line, Zeptosens is introducing a new revolutionary fluorescence-based detection technology for nucleic acid and protein microarrays. This will set new standards in microarray readout performance in terms of detection limits, quantification and automation. The initial products will be tailored for gene expression monitoring, even in minute amount of tissues such as biopsies, and for multiple simultaneous determinations of immunoanalytes such as biomarkers. Typical application areas will be in the life science research and development.

Technology Basis: The current core detection technology is based on the planar waveguide principle, an approach backed by eight years of research and development by the founder team in an industrial environment. Zeptosens has access to a broad IP and know-how basis in optical detection technologies, nanotechnology, micro-fluidic designs, miniaturized bioassays, and surface chemistry.

Business Strategy: The strategy is to provide academic and industrial customers with complete bioanalytical solutions that will enable them to obtain information on critical parameters faster and with higher quality, therefore enabling them to make decisions more efficiently and more reliably. For marketing and service Zeptosens relys on the competencies of established partners.

Technology Intelligence system at Zeptosens

Zeptosens management is continuously concerned with the question "what is next?" Therefore, they run a formalized TI system. There are several TI elements that will be illustrated in this section: the "business developer," the "technology watch," and the "business strategy meeting."

Business developer: There is one person who is charged with scouting for "deals" in proteonics. Deals mean business opportunities, which may also be research collaborations. However, these collaborations are expected to result in turnover later on. Therefore business development means "is there a problem in search of a (Zeptosens) technology." Through these scouting activities, the business developer comes in contact with numerous scientists, and thus, with new technologies. Because of his scientific background in chemistry he is able to analyze technical information. Insight from the scouting activities is transferred to other employees informally and formally by means of official meetings, such as the weekly team meeting (cf. "technology watch").

Technology watch: The major Technology Intelligence element is an agenda item in the weekly team meeting. This agenda item is named "technology watch." The technology watch mainly deals with one question: "are there potential opportunities and threats in the technological environment?" To answer this question, each employee contributes with knowledge that she/he could build on during her/his daily work. Some employees have specific tasks. First of all, because of the very nature of his work, the business developer is one of the "main actors" in the technology watch. In addition, there is one person working 100% on patents, which also makes him a specialist in knowledge about new technologies. Furthermore, other employees have formal tasks, which may be as a project leader in research collaboration, or as a direct link to a partner firm. However, other employees also contribute to the technology watch with knowledge from periodicals, internet searches, informal networks etc.

In this way, each employee has an entrepreneurial responsibility. An enormous advantage of such a "forced" exchange of trends about the technological environment and the emerging ideas is the increased pace with which to act on them, and the increased organizational knowledge base, and thus, potential innovations.

Business strategy meeting: Another TI element is the monthly business strategy meeting. In fact, insight from the technology watch is processed, and strategic decisions are made based on the generated intelligence. A particular point is the fact that next to the CEO and the heads of the teams, any interested employee may participate in this meeting. The only condition is that participants must also contribute actively to the meeting. This makes business strategy a democratic affair.

System overview: These elements together build a TI system, which is illustrated in Figure 7.19. In fact, there is a defined process that allows Zeptosens to be informed about trends in the technological environment. The process can be described by the value creating scheme. The institutionalized agenda item and the scouting mission of the business developer represent TI mission and goals. Some roles are explicitly allocated, other activities are executed informally and democratically by all employees. This system reminds us of the gatekeeper organization at Straumann. A particular characteristic is that top management is fully integrated in all TI activities, which means that any employee, if interested and competent, may participate in decision-making. Analysis is supported by certain methods, such as regressions.

Comparison of Technology Intelligence in start-ups and at Straumann

Observation of some start-ups gave insight into how they are organized to obtain information about facts and trends in the technological environment, and who fulfills these activities. Based on these observations, a comparison between the most important characteristics of TI at start-ups and at medium-sized companies, such as Straumann, gives insight into the validity of the solution at Straumann for start-ups. By the nature of this vertical validation case, the discussion below differs from comparison sections in the horizontal validation cases.

Need for Technology Intelligence

The need for Technology Intelligence seems to vary between start-ups, and therefore is not always comparable to needs at Straumann. The main difference may be companies' age. While Straumann has developed, produced and marketed products for decades, start-ups are in the early stage of company history. In this stage, the very nature of the start-up is usually to have one business idea, which is normally based on one pacemaker or key technology. Therefore, the start-up company is well informed about the actual state-of-the-art technology, as well as about trends in this technology field. Some start-ups therefore do not feel that intelligence activities are necessary.

However, some start-ups express their need for intelligence activities through an informal or formal TI process. The small sample does not allow a clear statement to be made, but the tendency is that the larger the start-up is, the more intelligence is performed systematically and shared among the most competent employees.

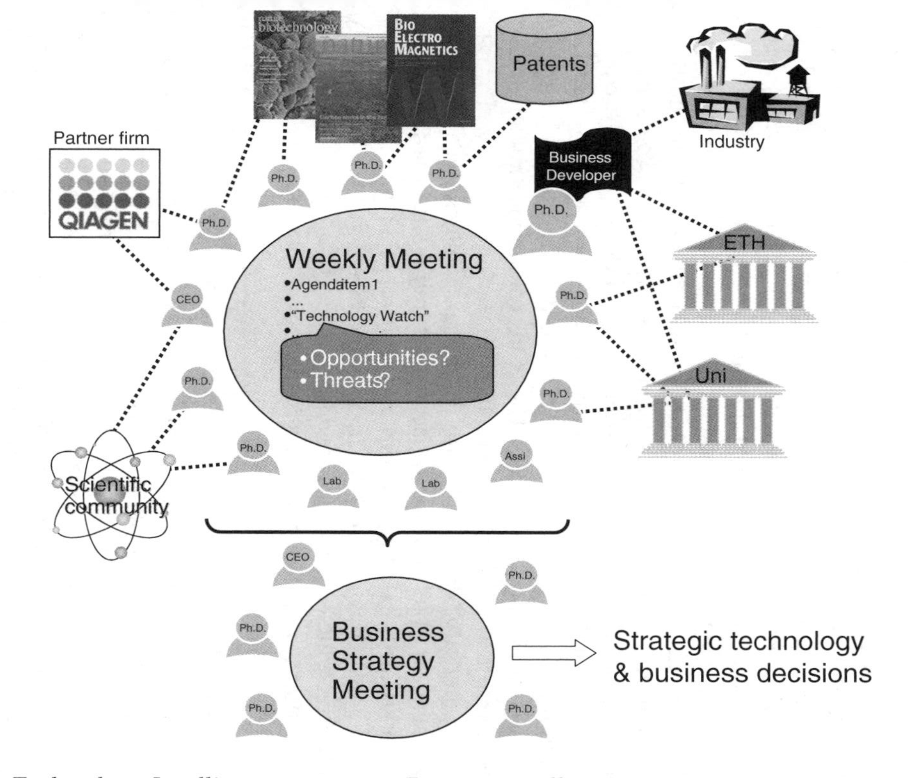

Figure 7.19 The Technology Intelligence system at Zeptosens – illustrative

Gatekeeper organization

This task-sharing organization corresponds to the gatekeeper organization, i.e. the Opportunity Landscape idea, at Straumann. The difference is that in start-ups, issues are not mapped such as in the Opportunity Landscape. In fact, the main goal of the institutionalization and the visualization in the Opportunity Landscape is to assure transparency. This does not seem to be, a priori, necessary in very small companies because there is a good overview of people and activities. Nevertheless, the Opportunity Landscape could be interesting for start-ups anyway, because since most activities have already been done, institutionalizing the Opportunity Landscape would mean giving the activities a face without additional efforts. Next, for internal purposes the Opportunity Landscape may be useful for external communication, for example to illustrate the company's knowledge base to investors. The additional dimension of the Opportunity Landscape furthermore could indicate missing or yet to be built competencies. In addition, the participative approach of the Opportunity Landscape in a start-ups means that decision-makers, in particular the CEO, are directly involved; they are usually gatekeepers themselves. Therefore, communication and knowledge transfer does not cause a problem in start-ups.

In summary, the Opportunity Landscape seems to be a promising element for TI in start-ups. In fact, this is in line with the TI approach which already exists in start-ups. As is true for Straumann, defined as a medium-sized company, no start-up could be observed to be organized industry-wide or with other companies for TI purposes. This emphasizes that TI effectively is an internal concern that cannot be cooperatively done among companies.

The use of Technology Intelligence sources and methods

Information sources and TI methods are the most tangible elements of a TI system, and they can be directly compared.

There does not seem to be a difference between **information sources** at Straumann and TI executing start-ups. In both cases, the prime information sources are informal expert networks. Start-ups are particularly linked with universities or large companies because several of them were born by means of a spin-off. Since employees are mostly scientists, periodicals are read regularly, and congresses and trade fairs are frequently attended.

Also the use of **TI methods** does not differ between the observed start-ups and Straumann. They both are rather hesitant in the use of "technical" methods. Besides some basic methods, such as an ordinary time-series regression, they generally adopt multidisciplinary, intersubjective, opinion-based methods.

In conclusion, both start-ups and medium-sized companies seem to have similar conditions to obtain and work on information.

Vertical validation case D: large companies

This second vertical validation case aims to discuss the generated Technology Intelligence elements at Straumann in comparison to TI systems at large, multinational companies. Recent research shows that the organization of TI in large companies depends on the competitive environment, but also on internal characteristics, such as R&D organization (centralized vs. decentralized), decision-making competencies and processes (centralized vs. decentralized), company culture, technology, and application competencies, etc. Therefore, two large, multinational companies that differ in most characteristics build the basis for this comparison. The described companies are Novartis and Daimler-Benz. Daimler-Benz merged with Chrysler in 1998. Discussion in this section still refers to the Daimler-Benz before the merger. In general, information reflects the status of both companies in 1997. Information with regard to these two companies originates from the empirical base of two earlier studies at the ETH Center for Enterprise Science, i.e. Lang (1998) and Lichtenthaler (2000).

First, the two companies will be presented on the basis of these recent dissertations. This presentation contains the company profile, R&D and technology management organization and, with most emphasis, elements of TI. However, for a more detailed description it is recommended to refer to the original works. Then, TI at these two companies will be compared to the solution at Straumann.

Technology Intelligence at Novartis

Novartis stands exemplarily for a large, multinational company in the science-based life science industry.

Novartis: company profile

Novartis is a globally active company in the life science industry. In 1997, about 90,000 employees contributed to US$21 billion net sales. The company is divided into three divisions, which are "healthcare" (60% of net sales), "agribusiness" (27%), and "nutrition" (13%). Each of these divisions contains sectors (Figure 7.20).

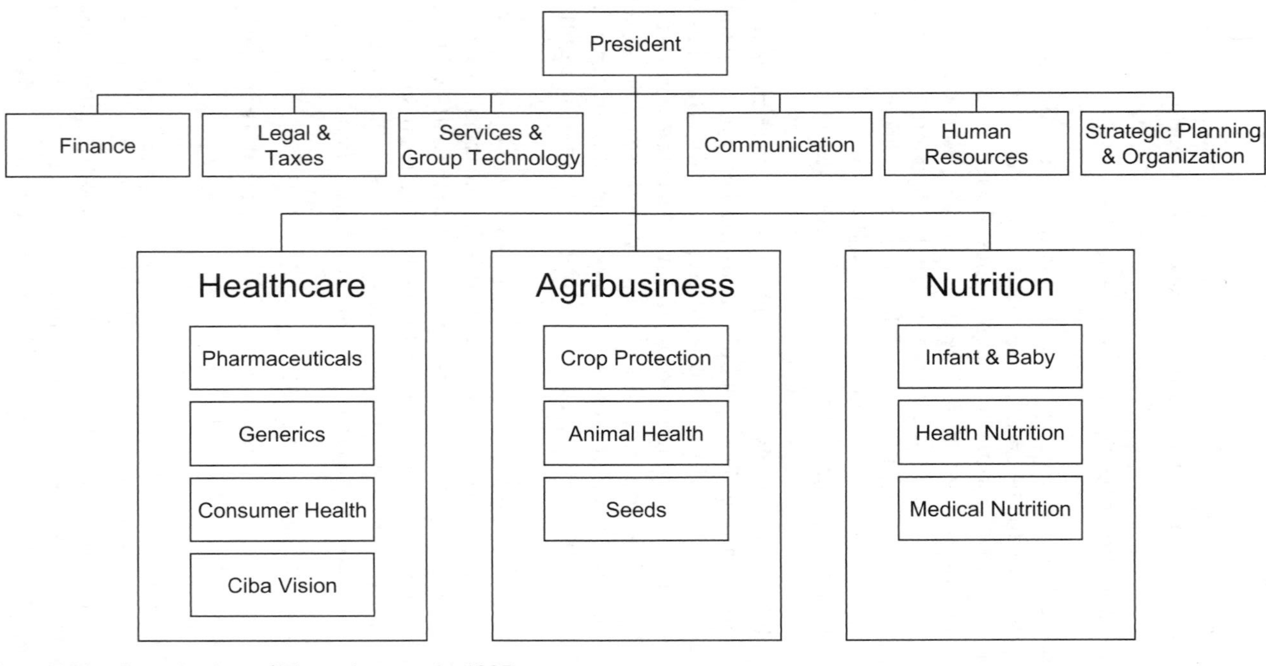

Figure 7.20 Organization of Novartis group in 1997

The group's mission is:

> *to have a positive impact on lives and to discover, develop and successfully market innovative products to cure diseases and enhance the quality of life.*

Within the healthcare division, Novartis Pharma is the most important sector with about 44% (US$9.3 billion) of the group's net sales. This is also the most technology-intensive sector with about 19% (US$1.8 billion) spent on R&D, compared to the group's average of 11.8% (US$2.5 billion). Because of the pharmaceutical sector's dominance and above average technology intensity, the focus in this section is on this sector.

The average development time of a pharmaceutical product is about seven to eight years. The goal of Novartis Pharma is to successfully bring to market three innovative products per year. The product pipeline of Novartis Pharma with 75 projects (24 in phase I, 14 in phase II, and 35 in phase III/registration process) seems to be a promising way to achieve this goal.

R&D and technology management at Novartis Pharma

R&D activities are fully decentralized to sectors. Exemplarily, R&D at Novartis Pharma is organized by a matrix of three dimensions: therapeutic areas, core technologies and senior experts. **Therapeutic areas** contain technological and scientific disciplines (for example dermatology, immunology etc.) which, in turn, are overlaid with a project organization. Coordination between therapeutic areas is managed by the monthly held therapeutic area board meeting that includes discipline directors, large project leaders and the head of therapeutic areas. **Core technologies** are technologies that are very important for their own business. Currently there are about ten such core technologies defined, which generally span the therapeutic areas. Coordination is assured by the monthly held core technology board meeting with core technology directors and the head of core technologies. The heads of therapeutic areas and core technologies are members of the research board. The research board is completed with **senior experts**, actually three of them, who directly support the head of research. Their tasks are to coordinate technologies between therapeutic areas and to explore new technologies and trends.

Responsible for general coordination of these three R&D dimension is the monthly-meeting Research Management Board (RMB), that consists of heads of therapeutic areas, the heads of core technologies, the senior experts, and the head of research. Project decisions, in particular rough resource allocation, are the main contents of this board meeting. However, therapeutic areas and core technologies are quite independent in strategic planning and "detailed" resource allocation. Thus, this is a bottom-up process that is "controlled" by an upper level.

R&D activities within these three dimensions are dispersed among several global research centers, mainly in Europe and the USA. Next to internal R&D activities, Novartis Pharma maintains a cooperation network. About 80% (US$1.45 billion) of R&D resources are spent internally, 20% (US$0.35 billion) for these external activities.

At the group level there is a central "Group Technology" that coordinates the sector's technology strategies, detects synergy potentials, is responsible for knowledge management and generation of new businesses. This group is supported by two scientific advisory boards, the Research Advisory Board (RAB) for product technology purposes and the Technology Advisory Board (TAB) for process technology purposes, which include, among other members, the most important heads of R&D and production.

In summary, the most important characteristics of R&D and technology management are:

- decentralized sectors and therefore very international R&D
- project are financed by sectors, some budgets for new technologies at the group level
- participative planning in sectors, coordination of synergies at the group level
- science-based and technology-driven business
- market-driven innovation culture and bottom-up-driven decision-making culture.

Technology Intelligence at Novartis: elements

Technology Intelligence activities are executed at the group's level and within sectors. At the group's level there is the so-called Future Watch, that is coordinated and animated by the Group Technology, and the Scientific Services. At the sector's level, numerous TI elements could be observed (Figure 7.21). The most important elements, as well as the Technology Group and the Scientific Services, are illustrated very briefly below. Again, the Novartis Pharma sector is chosen as example.

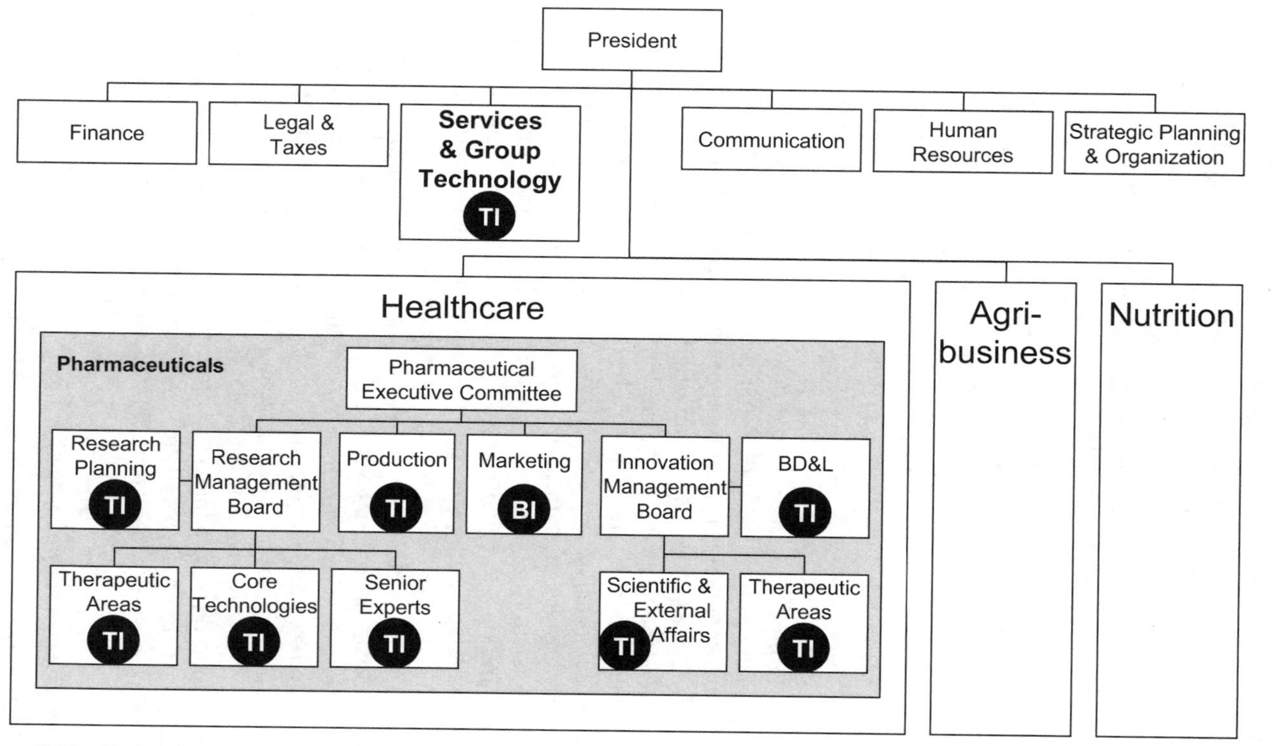

Figure 7.21 Technology Intelligence elements in Novartis Pharma

Future Watch and group technology (group level)

In order to scan and assess new technologies, Novartis installed the so-called Future Watch at the group level:

> *The main focus is to identify timely technological discontinuities that could be of interest to Novartis.*

The Future Watch is a participative, multidisciplinary and virtual monitoring tool that is animated by the Group Technology. About 150 voluntary scouts from all sectors build a gatekeeper network. They come together physically once a year for a meeting. The goal of this meeting is to initiate new scouts and to build mutual trust. However, the effective Future Watch meetings are organized via the internet. On a Lotus Notes basis, a group of scouts "meets" over a period of about two weeks virtually in order to discuss the evolution of a defined issue. These issues are proposed and selected democratically in advance. Next to individually proposed issues, other issues are identified by means of scenarios. Based on this definition of the observation area, each scout collects information. Most important information sources are the scouts' (= "gatekeepers") formal and informal networks. In addition to these networks they are supported by their R&D units. Hence, technological trends are assessed by a scout group. If the result is of strategic interest, insight is communicated to the concerned people. The primary client is top management. The Group Technology, which animates (and therefore participates in) the Future Watch, communicates this generated intelligence directly or after another assessment by the scientific advisory boards to top management.

Since this is a distinctive bottom-up process, that in turn is supported by important people by means of the scientific advisory boards, acceptance of the generated intelligence is high, which facilitates transfer into the various sectors. The fact that most projects are financed directly by the sectors could be a hindering factor for adoption of a new technology because of the risk of a new technology. Therefore the Group Technology has at their disposal a budget in order to initiate a technology project. Thus, the Future Watch also builds a basis for resource allocation.

The process and structures of the Future Watch are depicted in Figure 7.22.

Scientific Services (group level)

Scientific Services is a profit center at the group level. They provide services for Technology Intelligence, such as database research to

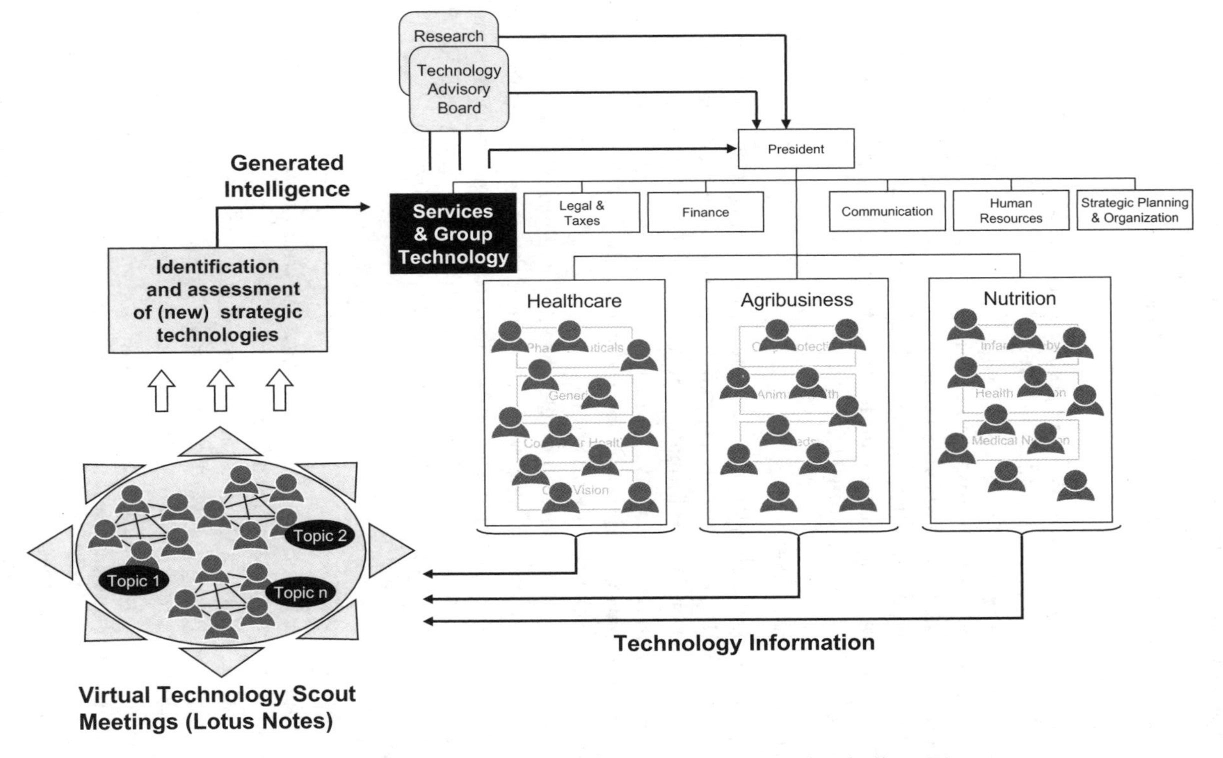

Figure 7.22 World-wide Technology Intelligence (Future Watch) at Novartis group level, illustrative

internal clients, i.e. in sectors. Clients clearly formulate a problem that the Scientific Services solve. This may be technology and/or competitor analysis. In fact, they deliver information rather than intelligence, i.e. analyzed information, the latter is done by the clients themselves.

TI by R&D employees

Technology Intelligence at the sectors level is considered a task of each R&D employee. A very important element of TI is identification and recruitment of outstanding junior scientists. Because they are considered more valuable information sources than an external expert network, junior scientists are engaged for a defined period, and then replaced by other junior scientists. The description of the following TI elements should therefore be understood from this point of view. They are, in fact, TI coordinating elements.

TI in core technologies

Exploitation of new technological facts and trends is a basic task of "core technologies." There is a systematized gatekeeper network within each core technology. Information need is formulated by the core technologies management board and the head of core technologies. The main information sources are periodicals, publications of start-ups, conferences, scientific reports from universities and an internal expert network. Next to information collection, the expert network is important for assessment. This network meets monthly. Communication of insight is through participation in the core technologies boards, semiannual research reports, colloquia several times a year and informal meetings. Communication for this network is also supported by an IT system.

TI of senior experts

The main tasks of senior experts are to coordinate technologies between therapeutic areas and to explore new technologies and trends. The focus of observation is on trends at universities and start-ups. Recent trends are analyzed at monthly committee meetings. Most promising trends can be validated in a laboratory that is attached to each senior expert. Since senior experts are on the research management board, communication of relevant insight is very direct.

TI in therapeutic areas

There are Technology Intelligence units for each therapeutic area. About three employees support the decision-making and planning processes of each therapeutic area. Through participation in these processes, information need is determined. Information is gathered from databases and internal networks. The latter are also contacted for assessment. Results are directly communicated in the mentioned processes or if necessary, directly to a concerned person.

TI in research planning

The staff function of "research planning" coordinates the resource allocation process for internal research, supports strategic planning across functions and coordinates external technology planning. Research project proposals that are submitted to the research management board pass through the research planning staff for an initial evaluation. At the request of the head of research, this group also analyzes particular technological and market trends. Since this group does not have sound know-how for information research, they often revert to "scientific services." Next to these requested activities, a regular product of the research planning group is a compilation of reports on the status of actual research projects and relevant trends within specific therapeutic areas.

Another task of this group is to identify and assess potential cooperation partners, which are usually universities and start-ups. This task is accomplished by two persons who build the "technology acquisition intelligence" unit. Most cooperation partners are searched for proactively and are directly identified by the researchers themselves. Thus, the main task of the technology acquisition intelligence unit is to organize the administrative aspects of these joint operations. However, numerous cooperation proposals come directly from potential partners. The technology acquisition intelligence unit is the centralized contact point at Novartis Pharma. This unit then organizes an internal assessment that reverts to a multidisciplinary expert network. Next to these reactive tasks, the technology acquisition intelligence unit also fulfills, to a certain extent, systematic monitoring of start-ups in which Novartis is actually not yet active.

Business Development and Licensing (BD&L)

Parallel to the technology acquisition intelligence unit, the Business Development and Licensing unit is focused on initiation and transaction of strategic joint operations in the pre-clinical phase. The goal is to close gaps in the product pipeline. Therefore this unit, employing 35 people, is a service unit to therapeutic areas and core technologies. They get information through commercial databases of technology, product and company information, as well as through continuously updated internal databases. Unfortunately the databases are not coordinated with the technology acquisition intelligence unit's databases.

Scientific and External Affairs

The Scientific and External Affairs unit is a scanning unit that has been created to benefit from the extensive expert network of the former head of research. This unit employs three people who scan for application technologies.

TI in production launch

Implementation and validation of new production technologies takes two to three years. Since this is very resource intensive, new process technologies are mainly implemented for new products. (Re-engineering of the production process of a recent product by implementing a new technology is only interesting if costs can be reduced by 20–30%.) Because of the long validation time, the focus is on mature technologies with a low uncertainty. TI in production is not formalized but a task of each process researcher.

Holistic view of Technology Intelligence at Novartis

The character of the Technology Intelligence system at Novartis reflects the decentralized character of R&D and the autonomous character of sectors. The goal of the Future Watch is to explore new businesses and observe facts and trends across sectors. This covers well the focus of the "looking-beyond area" in Figure 4.4. In turn, at the sectors level the focus is on specific business and copes therefore with the "keeping-abreast area." A close interaction between these various levels is not institutionalized.

Technology Intelligence at Daimler-Benz

Daimler-Benz is representative as a large, multinational company in the automotive industry.

Daimler-Benz: company profile

Daimler-Benz is one of the world's leading automotive, transportation and services companies. In 1997, Daimler-Benz turned over more than US$ 69billion and employed more than 300,000 people world-wide. There are 23 business units in four divisions: "passenger cars" (40% of turnover), "commercial vehicles" (30%), "aerospace" (12%), "services" (12%), as well as in the directly managed industrial business units, "rail systems," "microelectronics," and "diesel engines" (6%).

Daimler-Benz's main vehicle lines are luxury and high-class passenger cars, off-road vehicles, sports cars, and commercial vehicles. In 1997, a total of 715,000 passenger cars and 417,000 commercial vehicles were sold.

The group's mission is:

to develop, to manufacture and to sell the world's leading automobiles.

Daimler-Benz has known a number of reorganizations since the merger of two German car manufacturers, Daimler and Benz in 1926. In the 1980s and early 1990s Daimler-Benz acquired several industrial companies in order to establish itself as a major "integrated technology group" with technological foundations in automobiles, aerospace, microelectronics, and defense systems. Some of these businesses were sold later, when under new leadership Daimler-Benz began to concentrate on specific competencies and competitive strengths. R&D units that had been acquired in the wake of the diversification were brought under centralized control. They remained with Daimler-Benz's R&D after their former parent companies were divested.

R&D and technology management at Daimler-Benz

Daimler-Benz presents one of the highest R&D budgets worldwide. Total R&D expenditures in 1997 amounted to US$5.4 billion, which results in a technology intensity of 7.9%. About three quarters of R&D is vehicle-oriented, supporting the main businesses of Daimler-Benz.

Daimler-Benz does not heavily engage in basic research. Daimler-Benz's R&D includes product development, advanced development, and research. Research and development are clearly separated from an organizational point of view as well as by its underlying strategies and missions. Development is almost exclusively carried out by business units and amounts to US$5.1 billion. Mid- and long-term-oriented research, in turn, is carried out at the group level in the centralized Research & Technology (R&T) department. In 1997, R&T had a budget of US$280 million (0.4% of turnover). Research activities are increasingly financed by business units (about 50% of projects), but selected long-term projects which are of importance to more than one business unit and which are also of strategic importance for the future competitiveness of the Daimler-Benz group tend to be financed by corporate funds. Examples of such projects are multimedia, satellite based navigation and communication, and environmental production processes. This is in line with the long-term orientation of strategic technology planning with a time-horizon of about ten to fifteen years, while the group's strategy focused on three to five years.

R&T is organized in four research units and five staff positions (Figure 7.23). Again, the main research units follow their historic origin. For instance, the first R&T unit originates from the former Daimler-Benz research institute and is still mainly active in automotive research. Each research unit is divided into several laboratories and led by a director who reports directly to the Chief Technology Officer (CTO).

The research laboratories are mainly based in Germany with the most important site in Stuttgart. In total there are about 1,400 researchers and 800 temporary researchers attached to R&T in Germany. In addition to the German sites, in the 1990s four international research sites in Portland (USA), Palo Alto (USA), Shanghai (China), and Bangalore (India) and two listening posts in Moscow (Russia) and Tokyo (Japan) were established. In fact, the international sites are labeled differently: Research lab, transfer lab, research center, R&T center. All of them are understood as research sites in this book. These international sites employ less than 100 researchers, which underlines the very centralized R&D character.

The business system of R&T consists of three dimensions representing the customer, research product, and technology. **The customer** represents all business units of Daimler-Benz, their technology strategies and research programs. On the basis of their technology strategies, R&T defines customer-oriented research programs. For this, eleven research program managers support the four research directors staying in close contact with internal customers. **Research product** stands for all projects conducted within R&T. For each project not only is the targeted scientific result defined but also the added value it might provide to the final customer in at least one business unit is determined. Each project falls into one of the 20 core research areas. Finally, **technology** represents the technological and scientific know-how of R&T. In order to fulfill all tasks, competencies in defined fields of technology have to be gained and maintained. The laboratories which belong to one of the four departments of central R&T, serve as centers of competence for different fields of technology and are responsible for technology monitoring, networking and benchmarking.

In conclusion, the most important characteristics of R&D and technology management are:

- centralized and therefore low-intensity of international R&D
- long-term projects financed by the group, product development financed by business units

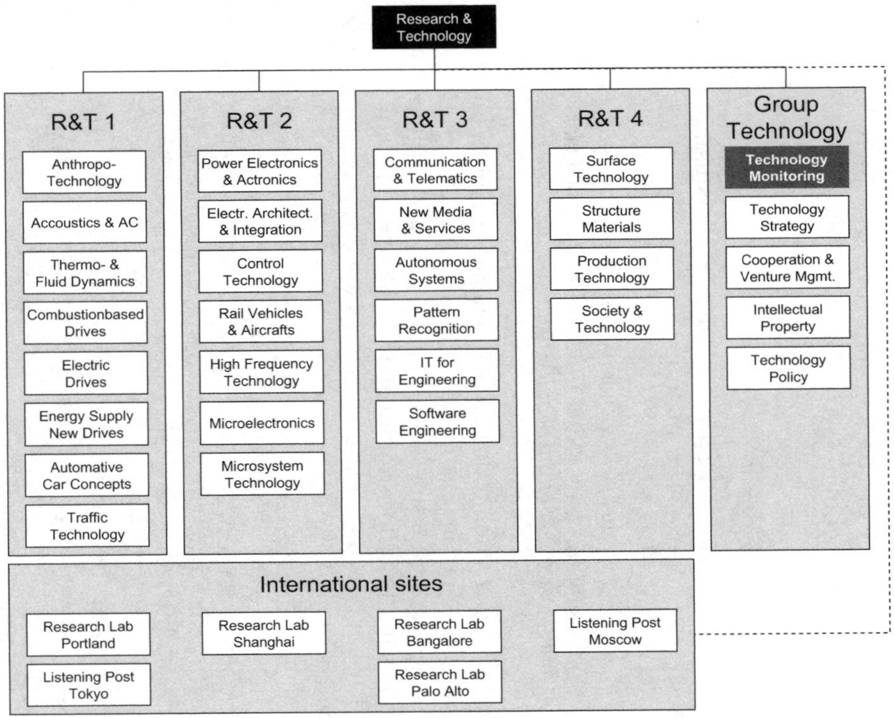

Figure 7.23 Organization of the research and technology department at Daimler-Benz

- participative planning at the group and business unit level
- technology-based and market-driven business
- science and market-driven innovation culture and business unit oriented decision-making culture.

Technology Intelligence at Daimler-Benz: elements

Technology Intelligence activities are mainly executed at the group's level, some activities in business units cover TI aspects. The most important elements of the TI system at Daimler-Benz are depicted in Figure 7.24.

Technology Monitoring group

The Technology Monitoring group is the most important centralized Technology Intelligence element. Its mission is:

> *Technology Monitoring at the Daimler-Benz group is a process that is managed centrally, which informs the S&T board, technology management and research and business units about external technologies.*

The Technology Monitoring group therefore sets up an intranet IT solution. This solution functions as follows: Firstly, a huge amount of relevant information is stored in a database with an "intelligent" scanning software that allows information to be retrieved with a search tool. Thus, information is stored centrally and can be retrieved on demand decentrally by decision-makers. Typical decision-makers are top and middle management, who at the same time are experts in their field. The goal is that information is simply provided, for example by information brokers such as "Pointcast," and that the user himself assesses the information. In parallel to this information-pull, information is also pushed to concerned people in the function of a user profile.

Next to this information delivering activity, the Technology Monitoring group has organization and coordination tasks. The group is the depot for information about technological facts and trends that do not fit within the above mentioned "informal" process. This information is then assessed by technology colloquia. Then, the group supports initiation of TI activities within laboratories and large research projects. Furthermore, the group is the competence center for TI methods.

Important tasks of the Technology Monitoring group are represented in Figure 7.25.

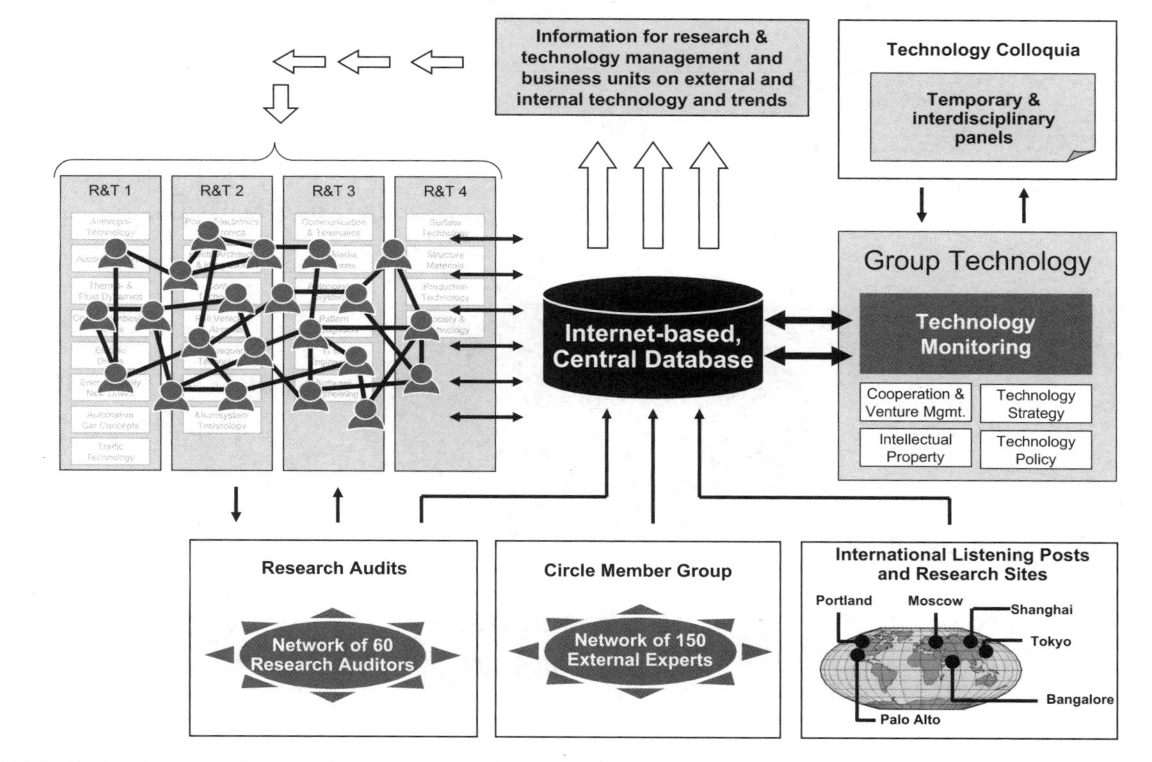

Figure 7.24 Technology Intelligence elements at Daimler-Benz, illustrative

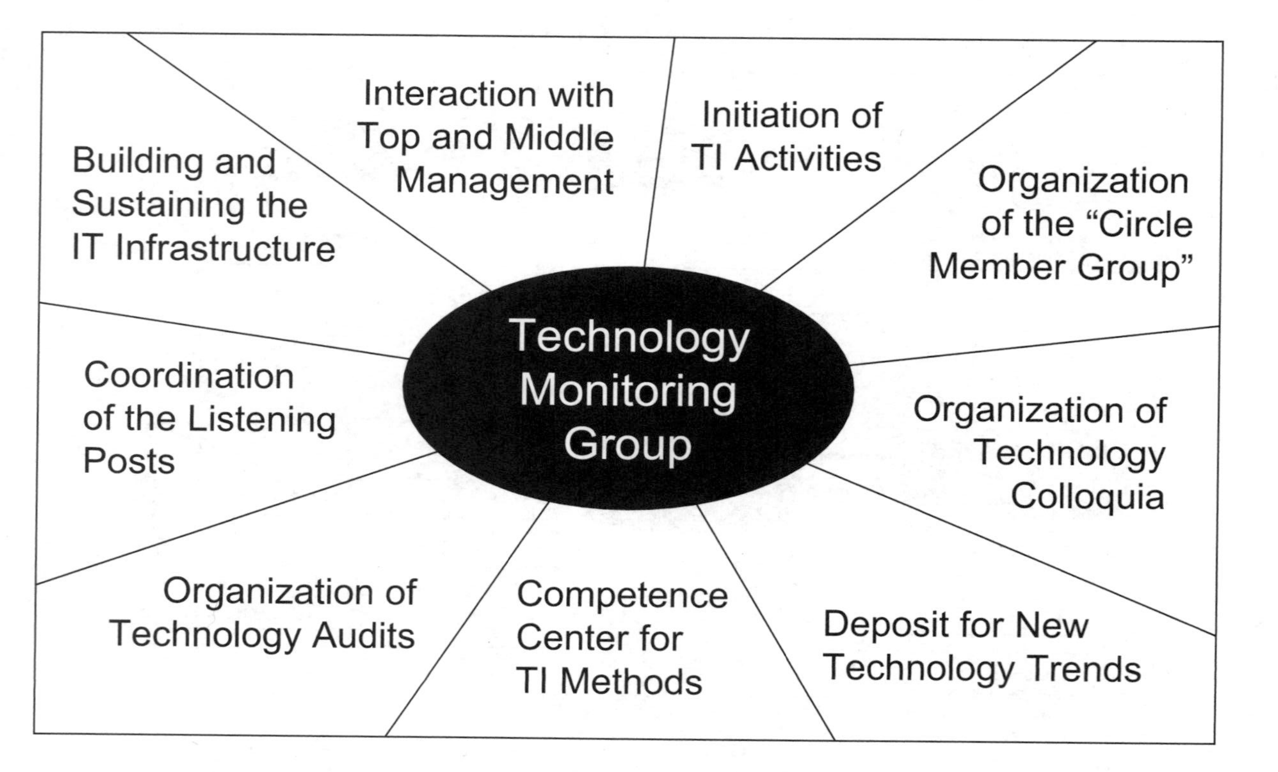

Figure 7.25 Main tasks of the Technology Monitoring group

Listening Posts network and international research sites

Two listening post were established in Tokyo and Moscow. Small teams are responsible for observation of key technologies areas and identification of new trends, establishing research joint operations, establishing and maintaining expert networks, and technology and market intelligence services. These listening post are tightly managed by the Technology Monitoring group. The biggest challenge is to coordinate information need with observation activities and information sources, in order to avoid an information overload.

The **Tokyo** Listening Post was established in 1990 to find out about Japanese automobile technology. In the beginning ten people were employed: three Germans and seven Japanese. The latter are important in order to have access to networks and to understand information within the cultural context. Germans, in turn, transfer information in the company context. The team's job description lists basic monitoring and project work. In order to cover a large number of technology fields and organizations each member of the unit specializes in a specific technology field and works on building his specific network. Basic monitoring consists of keeping track of daily life in Japan: Newspapers, press conferences, annual reports, research reports and conferences are monitored. The information gathered is translated and put into a newsletter which is structured according to technology fields. This newsletter is sent to Stuttgart in order to reach decision-makers. Next to the basic monitoring task, "clients" from research can also trigger an intelligence project, i.e. to identify key research institutions or to facilitate a visit by experts. The cost of basic monitoring in this listening post, which amounts to about US$1.35 million, is covered by Daimler-Benz" headquarter.

Due to the opening of Eastern Europe and several visits to high technology centers within the former Soviet Union, in 1993 Daimler-Benz also established a listening post in **Moscow** with employees (one German and five Russians) who try to establish contacts to research institutions and companies involved with material sciences, information sciences, and mathematics. The results of these activities vary substantially. The group has encountered difficult conditions: A research institution might open its doors to visitors one week and block any overtures the next. A contact once established therefore might not last very long. Furthermore, common information sources are often not fruitful because former Soviet scientists followed a publication and patent culture other than expected.

Four research sites were opened in Portland, Palo Alto, Shanghai and Bangalore in order to keep track of emerging technology developments world-wide.

In 1995, a research and technology center was established in **Palo Alto**. The mission of the center is to keep Daimler-Benz up-to-date about the accelerating revolution in transportation, communications, and related new technologies and services. Originally designed as a technology listening post, it was enlarged to a self-supporting research center as a critical mass and competency was necessary to understand and absorb the technological innovation in the surrounding area. One of its most important tasks is to facilitate intellectual exchange between US universities and research institutes and Daimler-Benz researchers around the world. A newsletter resembling the Japanese example has been compiled. Since language barriers do not cause problems, and since most information is also available on the internet, the value added from this newsletter is not information itself but the selection of information having in mind the development of the Silicon Valley region.

The **Portland** vehicle system technology center was established in 1996 to support a Daimler-Benz subsidiary with the early integration of specific technology and market know-how into their product development. Moreover, short organizational and geographic distances facilitate the cooperation with leading key partners and scientific institutes engaged in system modeling and simulation research. Another focus is on human factors engineering: The evaluation and optimization of manipulating technical instruments is crucial to produce user-friendly car controls. The Portland center closely cooperates with vehicle research units in Germany and other international R&D teams.

The **Shanghai** center was established in 1995 and supports a Daimler-Benz subsidiary which is engaged in microelectronics. However, the initial goal of this center was to enable production in China.

In 1997, Daimler-Benz established another research center in a crucial location for information technology and communications research where exceptional scientific progress in the software sector is being made, namely in **Bangalore**. Also, the main goal is not to monitor but to benefit from the cost advantage of software development in India.

Circle Member group

The Circle Member Group (CMG) is an information network consisting of distinguished scientists from all over the world. The know-how flows from top scientists into Daimler-Benz's research departments and its researchers can keep abreast with the latest research results and applications world-wide. About 150 leading scientists were selected for the circle member group. They were selected from a broad range of disciplines, also ensuring a good regional mix. Besides a proven academic excellence, they are expected to keep track of international state-of-the-art developments, and bring along excellent contacts to other institutes worldwide.

Each CMG member has a mentor within Daimler-Benz, but dialogues and information exchange is on a personal basis. This does not constitute an exclusive consulting agreement: The main function of the CMG is the integration of top people from leading institutes who, in this way, also gain insight into Daimler-Benz's research activities and assess its R&D projects as experts in their field. This external expert network is coordinated by the Technology Monitoring group.

Two times a year CMG members meet to transfer the group's long-term strategy and to determine future disciplines of interest. At the same time, these meetings build a platform for a creative idea exchange.

Technology colloquia

These colloquia are interdisciplinary and flexible panels for technology and trend assessment. The Technology Monitoring group organizes these panels and invites numerous researchers from different research areas for an assessment of those trends which cannot be handled with the intranet intelligence solution. Attending these colloquia is on a voluntary basis.

Research audits

In order to get direct feedback concerning quality and competitiveness of Daimler-Benz's research activities, a network of more than 60 external research auditors was established. These experts evaluate projects or research fields. The research auditing system is an ambitious policy that applies total quality management principles to research. Firstly, external top-level scientists judge the quality of research work and compare it with the world-wide leading state-of-the-art technologies. Secondly, assessments by internal "customer" within business are taken into account.

Up to 5 research audits are carried out each year. Each audit is performed by a group of 10 experts, including 5 business managers from the Daimler-Benz manufacturing divisions representing customers and

5 external scientists serving as independent experts. Normally, between 20 and 30 researchers are involved in each audit. The research audit is geared to give Daimler-Benz research and division executives a feeling for where they stand in the international scientific community and what the demand is for new innovation. The basic idea is to bring together the people who can best analyze new science: The researchers themselves, the acknowledged international experts in the field and the business managers who will be commercializing it. The research audits are not a mathematical formula or a controlling instrument, but rather a process for discussing the international position in a selected area of research. During the two-day audit, the research program is evaluated according to technology uniqueness, market potential, expertise of researchers, and management strategy.

Exceeding the original intention at its conception, this network of auditors has in the meantime turned into a circle of first-class advisors and consultants for Daimler-Benz R&T.

TI in business units

In business units Technology Intelligence is a part of Business Intelligence. Business Intelligence units are short-term oriented and focus on competitive aspects. Scattered among all business units, about 40 business intelligence specialists are employed. For technological competitive aspects they work in close collaboration with the Technology Monitoring group.

Holistic view of Technology Intelligence at Daimler-Benz

The character of the Technology Intelligence system at Daimler-Benz reflects the centralized character of R&D in heterogeneous disciplines. The intelligence process involves researchers at German research sites, as well as researchers or representatives world-wide by means of international listening posts and research sites. Coordination and animation is guaranteed by a central Technology Monitoring group. An attribution of TI activities to the "keeping-abreast" and "looking-beyond areas" is not ostensible.

Comparison of Technology Intelligence at large companies with the solution at Straumann

The presentation of the Technology Intelligence system in two large, multinational companies showed different ways how the system can be organized. The most important insight is that this organization depends strongly on the organization of R&D and technology management, on the organization of the company's structure, and on the business itself. Novartis's sectors are very independent and R&D follows a very decentralized organization. The (short-term) TI activities fall within the sectors' responsibility; however, a cross-sectoral initiative, the Future Watch, also guarantees long-term awareness of trends in the technological environment. Daimler-Benz, in turn, is organized very centrally. The company also performs most TI activities centrally, assisted by some international research sites. However, TI activities in these sites are strictly coordinated by the headquarters' Technology Monitoring group.

Based on these observations, a comparison between the organization of a TI system in large, multinational companies and the TI system at Straumann, which represents a medium-sized, technology-based company, gives insight into common and differing characteristics. At the same time, the comparison serves to validate Straumann's solution also in relation to large companies. For this comparison, some characteristics and elements are discussed separately.

Characteristics of R&D and technology management

Because of their importance for Technology Intelligence, the characteristics of R&D and technology management in the two large companies are listed again and compared with the characteristics of Straumann (Table 7.2).

The main difference is the fact that Straumann's internal R&D is exclusively done centrally within the headquarters, which seems to be typical for an SME. The degree of internationalization of internal R&D is therefore very low. Some research collaboration with some worldwide located institutes makes the amount unequal to zero. However, funding of projects at Straumann is planned by headquarters. With the exception of some projects which are financed by the ITI, technology planning is a top-down process at Straumann, while in the two large companies this is a participative process. But in fact, this aspect does not seem to be comparable, because within sectors of the large companies, planning is comparable to Straumann.

Straumann's business is driven in the same manner as it is for Novartis. Since both companies are in the life science business, this is not surprising. This characteristic is enhanced at Straumann because of the shift from bio-mechanics to BioTech in oral implantology.

Finally, innovation culture at Straumann is comparable to Daimler-Benz. Again, as for planning, decision-making culture cannot be

Table 7.2 Common and different characteristics in R&D at Novartis, Daimler-Benz and Straumann

Aspect	*Novartis*	*Daimler-Benz*	*Straumann*
Organization of R&D and internationalization	Decentralized sectors, and therefore very international R&D	Centralized, and therefore low-intensity international R&D	Centralized, and therefore low-intensity international R&D
Funding of R&D	Project are financed by sectors, some budgets for new technologies at group level	Long-term projects financed by the group, product development financed by business units	No sectors, therefore all projects financed by the group.
Technology planning	Participative planning in sectors, coordination of synergies at group level	Participative planning at group and business unit level	Top-down long-term planning, participative product development planning
Business drivers	Science-based and technology-driven business	Technology-based and market-driven business	Science-based and technology-driven business
Culture	Market-driven innovation culture and bottom-up-driven decision-making culture	Science and market-driven innovation culture and business unit oriented decision-making culture	Science and market-driven innovation culture and top-down-driven decision-making culture

compared directly because the entire Straumann company is comparable to business units of the large companies.

Overall Technology Intelligence system characteristics

The previous sections showed that these large companies have formalized their Technology Intelligence system, and how it is organized. While TI elements at Daimler-Benz are mostly organized at the group level, they are delegated to sectors at Novartis. This fits with the organization of R&D and technology management. However, in both companies TI is a process that is realized participatively by numerous employees, and coordinated by a "technology" group. This process is understood as a learning process and a subprocess of technology management. The TI system in both companies supports technology planning and resource allocation processes.

Considering these companies, which are representative of large, multinational companies, the main characteristics of the TI systems are comparable with the solution that has been designed for Straumann. In particular one main core element at Straumann, the gatekeeper approach, can be observed in both large companies discussed. Therefore, characteristics of the gatekeeper approach will be discussed more in detail. The other core element at Straumann, the screening process, will also be discussed with regard to the large companies. Finally, other elements are only briefly compared.

Gatekeeper approach

The gatekeeper network, or the Opportunity Landscape, is one of the core TI elements at Straumann. This is also true for TI at the group level at Novartis. The Future Watch's scope fits with the scope of the Opportunity Landscape, which is mainly to cover the "looking-beyond" area. The gatekeeper approach at Daimler-Benz is not that explicit. But the intranet-based central database initiative and the Circle Member group also adopt the idea of making gatekeepers collect and analyze information about facts and trends in the technological environment.

While the overall intent and the approach is similar in all three companies, details differ considerably. First of all, **issue definition** follows three different logics. At Straumann, issues are defined in a workshop and are, to a certain extent, stable. At Novartis, issues emerge from a democratic process among about 150 scouts, and from scenarios. Issues at Novartis have a temporary character. Considering Daimler-Benz, "issues" are not at all in the intranet solution, while in the Circle Member group issues are implicitly defined by the choice of external experts and by communicating the long-term strategy.

Also the **gatekeeper definition** is different. At Straumann, gatekeepers are defined top-down, and the gatekeeper role is formalized. Thus, they are known to be a prime internal information source. At Novartis, to be a scout is voluntary and temporary. Therefore they are not necessarily known to be experts in their field, and their impact is reserved to activities within the Future Watch. At Daimler-Benz, experts in the intranet solution do not know of each other, in fact they are not always aware of being a member of a gatekeeper network. Experts in the Circle Member group are external and their knowledge is transferred to Daimler-Benz" research by means of an internal representative. This makes positions in the CMG explicit and formal.

The **gatekeeper "process,"** i.e. the intelligence value-adding process, follows different logics in these three approaches. At Straumann, "gatekeeping" is a permanent and obligatory task and therefore is daily work of the gatekeepers. They organize themselves to produce intelligence. Communication follows a push as well as a pull logic, which means that some communication is standardized (for example the annual report), and some communication is informal (for example to revert to the gatekeeper's knowledge). At Novartis, the gatekeeper process is project-based, and thus temporally limited. Other than at Straumann, issues are tackled in teams, which make insight more intersubjective. Communication of insight strictly follows a push logic. At Daimler-Benz, the intranet approach is not really a process: A broad base of information is provided on the intranet and "processed" individually by the user her-/ himself. The value added with the CMG is to enable access to know-how rather than to be an intelligence producing process.

Coaching and coordination of the gatekeeper process is comparable at all three companies. At all companies, the process is coordinated centrally by a technology management group. They guarantee in particular the link to strategic decision-making, i.e. technology planning and resource allocation, and they can detect redundancies.

The fact that some details of the organization of the gatekeeper approach may differ does not seem to depend on company size. In fact, the overall idea of making the most competent people collect, analyze and communicate relevant information about the company's technological environment seems to be valid for a company of any size. This lets us conclude that the gatekeeper approach is a very promising and in practice a widespread Technology Intelligence element. Since the Opportunity Landscape is a formalization of the gatekeeper approach for TI purpose, its vertical validation with regard to companies smaller and larger than Straumann seems to be confirmed.

Screening process

Comparison of the screening process, in particular the intelligence aspect of it, is not conclusive. In fact, the early stages of the innovation process were not explored in detail at Novartis and Daimler-Benz. However, while at Daimler-Benz no TI elements is designed to react on information that is brought from externals, at Novartis the "technology acquisition intelligence" unit in the Novartis Pharma's research planning function is designed for such proposals. Central registration and multidisciplinary assessment present common organization as is for Straumann's screening process. Thus, this organization of the screening process also seems to make sense to larger companies, which confirms its validity.

Other TI elements

There are also parallels between the two large companies and Straumann for TI elements other than core elements. In fact, in all companies, a technology management group coordinates TI activities, which have already been discussed above. Then, TI activities are linked to business mission and strategy. Furthermore, next to the internal expert network, i.e. the gatekeeper approach, a formal or informal network of external experts (for example Straumann: ITI, Novartis: scouts' informal networks, Daimler-Benz: Circle Member group) facilitates access to external information about facts and trends in the technological environment. Finally, an IT platform (Lotus Notes or an intranet solution) supports TI activities in all companies.

Conclusion of further validation of generated Technology Intelligence elements

The aim of Chapter 7 was to "validate" the solution for a Technology Intelligence system which was generated during action research in one medium-sized company. The validation strategy pointed in two directions. In the one direction, the core elements of the solution at Straumann, i.e. the Opportunity Landscape and the screening process, were tested "horizontally" one by one in two companies which are comparable to Straumann in terms of size and technology intensity. In the other direction, organization of TI systems in companies that are a priori not comparable to Straumann were observed, and then compared to the TI system at Straumann. Since both small, technology-based companies (start-ups) and large, technology-based companies (Novartis and Daimler-Benz) were considered, this validation direction was labeled "vertical."

The purpose of this section is to consolidate insight from the four horizontal and vertical validation cases. Since some concluding remarks considering each case have already be made within the previous sections, the emphasis of this section is on the fundamental question:

> *Is the concept for the Technology Intelligence system that was designed and implemented at one medium-sized company (Straumann) a solution that makes sense for other technology-based SMEs?*

The answer to this question is developed by briefly exploring several arguments. First of all, and probably most importantly, the validation cases underlined, again, the **real need for Technology Intelligence** in practice. This need even builds the basis for the validity of the generated solution. The fundamental need seems to be independent of the company's size. The vertical validation showed that most companies of all sizes are organized in or are intuitively running a system that allows awareness of facts and trends in the technological environment. Motivation for running TI activities seem to vary. At RIC, for example, the main goal of TI activities is to give input into the Technology Roadmapping process and thus, to manage technological competencies. At Esec, in turn, the goal of TI is to initiate new product development. In general, motivation for TI activities reflects R&D and technology strategy, which has been illustrated through the Novartis and Daimler-Benz cases.

Even if the degree of sophistication seems to increase the larger the company is, the fact that some companies of any size adopt a **systemic and systematic approach** validates the approach for any company, and therefore for technology-based SMEs. Systematic is not necessarily equal to very formal structures and processes. It is rather a question of awareness of performing Technology Intelligence activities. Being aware, and therefore giving activities a system, begins with formulating TI mission and goals in order to set an observation strategy. In most observed companies, these mission and goals were derived from business and/or R&D mission and goals. This, again, points out the dependency between the TI system and organization in R&D and technology and innovation management. Another crucial characteristic of a systematic approach seems to be coordination. While in smaller start-ups coordination does not seem to be a problem, it is a challenge the bigger and more diversified the company is. Like for Straumann, it is common to all solutions that TI activities are spread decentrally and coordinated by a central staff. This is in line with the shift of Technology Intelligence from "intelligence of the organization" to "organizational intelligence" that has been observed in recent TI research. Thus, the solution at Straumann adopts "a modern approach." In particular sharing efforts among existing employees reflects this approach and is promising for resource restricted SMEs because ideally no additional personnel is necessary. However, a solution that is free of charge does not seem to be realistic.

Considering the elements of the TI system at Straumann, all elements or comparable approaches could be found horizontally as well as vertically in other companies. This indicates that these **elements reflect practice in use** of Technology Intelligence, and thus are consistent with reality. One core element, the Opportunity Landscape, was even implemented almost in its original version in another company (RIC), which is a validation *per se*. But also the fundamentals of the Opportunity Landscape, i.e. the gatekeeper approach, could be found in all companies of any size. In addition, comparable solutions to the screening process could be designed and implemented (Esec), or simply found (Daimler-Benz) in other companies, which is another validity for the solution at Straumann. In fact, elements of the solution at Straumann seem to be transferable to other technology-based companies, in particular other SMEs.

Validation of the **completeness** of the TI solution at Straumann is not trivial because the overall solution with all elements could not be implemented in another technology-based SME. It seems that this does not make sense because of the individual character of companies, in particular different R&D and technology management strategies. However, the generated solution at Straumann responds, firstly, to the requirements that Straumann formulated, secondly, to the aims of conducting Technology Intelligence that literature suggests, and thirdly, again, the practice in use within other companies that perform TI systems, which refers again to the integrity of the solution at Straumann.

Returning to the question whether the concept for the Technology Intelligence system that was designed and implemented at one medium-sized company (Straumann) is a solution that makes sense for other technology-based SMEs, what is the answer?

> *Considering the previous arguments we can assume that, indeed, the solution at Straumann appears to be valid for other technology-based SMEs.*

This does not imply that this is the only solution that is possible, but a promising and approved solution. Recommendations for what principles technology-based SMEs should follow to design and implement a TI system will be presented in the "management principles" chapter. There, recommendation may go beyond the generated elements designed for Straumann.

8
Towards a New Set of Management Principles

The aim of this chapter is to make a major contribution towards closing the gap in practice. This gap represents a call from reality for management principles guiding the practitioner to define and implement a Technology Intelligence system in a technology-based SME. Therefore, this chapter is the practical answer to the initial question: "How could a Technology Intelligence system be designed for and implemented in a technology-based SME?"

The nature of this chapter is different from the character of the other chapters. While much of this book follows scientific and sound argumentation, this chapter is normative and hands-on. Therefore, the proposals in this chapter are not thoroughly argued, but quite straightforward. However, any suggestions are based on practical experience gained during a number of years of research and a sound theoretical background in the field of Technology Intelligence. This also implies that the principles do not strictly depict the elaborated solutions offered during action research and the validation cases. This would be contrary to the most important insight of recent research and of this study, namely a strong dependency between the organization of the TI system and the company's context. Since the context differs from one company to another, the management principles cannot be seen as recommendations on a very detailed level. Therefore, the management principles give general indications.

This chapter reflects the realistic situation of a technology-based SME that wants to take advantage of relevant information about technological facts and trends of the company's environment. What principles should be considered?

The following management principles seem to be promising for Technology Intelligence in technology-based SMEs. They are listed in an order that corresponds logically to the questions which a technology-based SME faces when setting up Technology Intelligence.

Each principle includes the principle itself, key benefits of this principle, and explanations. In these explanations the practitioner will also find some detailed suggestions. This discussion should be understood from a technology-based SME point-of-view. Therefore, key benefits and arguments are related to SMEs' contexts. However, some principles or parts of principles would be of interest to any company.

Principle 1: Company Size

Technology-based companies of *any* size should perform Technology Intelligence in order to take advantage of relevant information about facts and trends in their technological environment.

Key benefits:

- Small, medium-sized and large companies **improve** decision-making **quality** with Technology Intelligence.
- **Opportunities** from the technological environment can be **captured**.
- **Threats** from the technological environment can be **avoided**.

As discussed in the introductory chapter, the existing management literature – in particular technology and innovation management – points to technological change as being a fact, and at the same time a challenge that technology-based companies presently have to face. There are numerous studies and concepts offered to answer this challenge. Most studies, and therefore most concepts are done with and are practical for large companies. This is also true for Technology Intelligence, which is a topic that seems to be of a particular interest with regard to the technological change; designed Technology Intelligence solutions are seldom applicable for SMEs. However, the "practitioner's voice" showed that there is a need for Technology Intelligence concepts for SMEs. In this book, such concepts have been designed, implemented and validated. This has shown that basically, it is possible for SMEs to perform Technology Intelligence.

Since there is obviously a practical **need** for Technology Intelligence in large companies *and* SMEs, and since solutions are **possible** for both company types, this first principle is formulated as a basic statement that technology-based companies of *any* size **should** perform Technology Intelligence.

Principle 2: Major Pitfalls

> Technology Intelligence in SMEs faces two major pitfalls: the "anyhow-argument" and the belief that it is not a good value!

Key benefit:

- Knowing the pitfalls before starting a Technology Intelligence initiative helps to **avoid discrepancies** during design and implementation as well as afterwards.

It is the very purpose of Technology Intelligence to provide the most relevant information and arguments for rational and comprehensive decision-making. The "**anyhow-argument**" is probably the strongest "weapon" against rationalism. What is the "anyhow-argument"? This argument expresses decisions that are made contrary to any rational and logical argumentation. Technology Intelligence is supposed to deliver information, and therefore arguments, that can be analyzed exactly to the extent that is necessary for the specific decisions to be made. Whenever the decision made is contrary to the generated intelligence, there may be two reasons: Firstly, the decision-makers do no trust the generated intelligence, secondly, the intelligence is contrary to the decision-makers' own interests. In the first case, the intelligence program does not meet the "customers" requirements, and therefore should be adapted. (In order to avoid such a futile intelligence program, the consideration of the following principles is recommended!) In the second case, results of TI are basically adequate, but they do not have an impact because decision-makers apply the "anyhow-argument." If such situations are the rule, there is only one suggestion: A Technology Intelligence system should not be installed in this company! It would be a waste of resources. Certainly some intuitive and rough decisions cannot be avoided, and that is basically satisfactory. It seems that particularly in SMEs some entrepreneurs often decide intuitively. But if a company decides to "officially" perform Technology Intelligence, such intuitive decisions should at least not be in opposition to results found through intelligence activities. A company that decides to install a Technology Intelligence system absolutely should be aware of this pitfall.

Another major pitfall is the belief that Technology Intelligence does not need any **resources**. This is sometimes the wish of SMEs, because they are restricted in resources. Technology Intelligence definitively does need some resources: money, time, knowledge, etc. The challenge is then to design a system that requires a minimum of resources, and even may optimize other resource expenditures. The following principles aim to propose a resource-optimized system for SMEs. The practitioner should be aware that the Technology Intelligence system itself needs additional resources, but a good TI system helps a company save resources (or even allows additional benefits) because decision quality is improved! Any necessary investment in this direction when consciously considered by top management becomes a good value for the money.

Principle 3: Systemic and Systematic Approach

> Technology Intelligence should be managed in a systemic and systematic manner in order to improve decision-making quality.

Key benefits:

- A systemic approach of Technology Intelligence improves **transparency** and therefore assures completeness and avoids redundancies.
- Managing Technology Intelligence systematically allows a consistent **development** of all elements in order to respond to changes.
- Involved people are aware of being Technology Intelligence workers, which makes the system more **effective** and **efficient**.

The expression "**systemic**" does not necessarily imply a formalization of Technology Intelligence, because the degree of formalization determines how the task should be accomplished. In turn, this principle states that TI tasks should be approached in a holistic way. This implies that all elements at least should be considered, and then organized in an appropriate manner. With regard to this, the company should be attentive that some elements, for example determining information sources, are not over-emphasized, while other elements such as defining observations areas, are neglected. It is recommended that the following elements be considered: TI system management, TI mission and goals, TI structure, TI process, TI tools (Figure 4.2). In fact, the precise design of each aspect depends strongly on the company context, and therefore cannot be delivered as a "cooking recipe." However, the following principles and the examples in the cases give the practitioner valuable inspiration.

Managing the system means that the system should be designed/defined, directed and developed consciously, i.e. **systematically**. The difference between design and definition is that definition refers to an existing concept or parts of a concept, for example TI tools. In order to avoid re-inventing the wheel, such parts of concepts should be adopted and adapted from existing solutions, whenever possible. If such concepts are not useful to the company, which is possible because of differing contexts, they have to be newly generated. This is equal to designing the concept. Directing and developing the system emphasizes the fact that once the TI system is designed/defined and implemented, it has to be maintained. Maintenance includes proactive and reactive adaptations to "new" situations. These new situations may be strategy changes, radical changes in the company's environment (which should have been detected with the TI system), personal changes etc. In fact, the best system seems to be the one that allows permanent changes. However, frequency and extent of TI system change should be in line with the general attitude towards changes of management systems in the company.

The contribution of the systemic approach to the **quality** of Technology Intelligence is increased effectiveness and efficiency of TI activities. This performance measurement will be explored in Principle 10. Already now one can say that a system increases transparency of activities. This transparency is positive in order to avoid redundant activities and to assure that Technology Intelligence concerns are all taken into account.

Principle 4: Strategy Relatedness

The Technology Intelligence System should present mission and goals which are directly linked to business mission and strategy. TI mission and goals set the focus of and guide Technology Intelligence activities.

Key benefits:

- The direct link between TI mission and goals and business mission and strategy stands for **effective** TI activities.

- This link justifies TI activities and therefore **resource expenditures** on these activities are also **legitimized**.
- TI mission and goals **determine** the definition of other TI elements and **guide** daily TI activities.

Business mission and strategy is the **basis** of a Technology Intelligence system. Various strategic orientations, for example technology leader vs. follower, opportunity seeker vs. threat avoider, in-house driven vs. cooperation driven etc., determine different organizations of a TI system. We suggest the consideration of two areas: the "**keeping-abreast area**" and the "**looking-beyond area.**" The first is rather short-term oriented and the focus is on the existing business. The second goes beyond this temporal and business limitation. However, companies that look beyond a certain limit should also be active in the keeping-abreast area in order to stay competitive today. Therefore, they should follow a dual intelligence strategy. This implies different TI activities for these two observation areas. Examples are given by the "Screening Process" and the "Opportunity Landscape" at Straumann. The Screening Process focuses on near-future product development, while the Opportunity Landscape builds the basis for long-term future innovations. Nevertheless, all TI activities are strongly linked to business mission and strategy.

This link, and therefore a TI system that is likely to be effective, is particularly strong if **top management** is involved in the TI system; as client and/or as TI worker. This seems to be one of the most valuable opportunities of SMEs in comparison to large companies. If top management is strongly involved, TI activities promise to lead to the intended results. In addition to this positive effectivity, efficiency should be favorable because communication is very direct. During this study it could be observed that top management tends to be more involved in TI activities the smaller the company is. Therefore, for growing companies, the challenge is to find the right point in time to take measures if top management reduces their involvement.

While business mission and strategy apparently influences the TI system, the contrary is also true. Insight emerging from Technology Intelligence may influence tomorrow's business strategy. This requires a strategy process which is open to such emergent information, for example a dual strategy approach (Abell, 1997).

Principle 5: Cultural and Organizational Fit

> The solution for the Technology Intelligence System should be compatable with the company's culture and basic organizational characteristics.

Key benefits:

- The Technology Intelligence system is well **understood** and **accepted** by the employees.
- Potential **conflicts** are **minimized**.
- TI activities are in line with other business processes and therefore are more **effective** and **efficient**.

Recent studies have found that the central contingency factors of the organization of Technology Intelligence are the company's basic organizational characteristics and the culture (Lichtenthaler 2000: 354). Observations in this book supported this insight. With regard to these factors a polarization of two company types seems to make sense: the culture-oriented company (for example Straumann, RIC) and the process-oriented company (for example Esec). In fact, this typology has not been developed in this book before. This is a simple post-observation which gives a discussion framework for this principle. In the following these two types are exemplarily illustrated, and consequences for Technology Intelligence will be briefly discussed.

Process-oriented companies are characterized by distinctive processes and hierarchies. This implies a high degree of formalization of decision-making, resource allocation and innovation processes. Strategy is usually a top-down process that defines other processes, such as the Technology Intelligence process. Since these process-oriented companies tend to have clear strategic goals, for example diversification, growth etc., TI mission and goals can easily be determined. Because goals and processes are usually well described and guide daily business, it is recommended to explicitly express these TI mission and goals by means of clear definition of an observation focus. If required, the commitment to undirected observation activities should also be clearly expressed. Roles with regard to Technology Intelligence should also be assigned explicitly in order to make the process work. Communication should be based on formal information exchange by means of formal meetings which are coordinated with the decision-making process. Communication support with IT solutions are an option but not imperative. However, IT should be deployed for database purposes. In conclusion, in process-oriented companies, a **high degree of formalization** of the Technology Intelligence system is recommended.

A special characteristic of **culture-oriented** companies is cultures that focus on an open information exchange and hence create a uniform business understanding over all hierarchical and functional levels. Such a culture is usually found in companies with proportionally a higher number of scientists and engineers, which is also reflected in the composition of the management. Innovation does not follow strict processes with clear gates, i.e. decisions, but is a participative and evolutionary process. Resource allocation is seldom a clearly defined process with specific requirements. While strategy is less explicitly formulated than in process-oriented companies, the common business understanding gives the TI mission and goals a direction. Despite this implicit understanding, it is recommended to determine and formulate these mission and goals explicitly. Such as for the other company type, directed and undirected observation fields should be pointed out. Regardless of the fact that innovation and decision-making processes are rather participative in culture-oriented companies, TI activities should be assigned to people in order to stimulate responsibility. Taking advantage of the open and informal information exchange culture, communication is favorable. However, some formal events allow assessment and if needed, improve transparency of activities. In addition to transparency, and therefore need for access to the "right" people, an IT solution for database purpose seems to make sense. In conclusion, in culture-oriented companies, it is recommended to **not over-formalize** the Technology Intelligence system.

SMEs tend to be culture-oriented companies. This is particularly true for companies in which top management is still highly involved in operational activities. Such an involvement mostly implies participative processes and informal communication. Therefore, attention should be paid that the degree of formalization of the Technology Intelligence system does not rival the relatively low degree of formalization of other processes. However, some SMEs present rather formal structures and processes. In this case, the Technology Intelligence system should be designed/defined as closely as possible to these processes.

Principle 6: Parallel Definition and Implementation

Implementation of the Technology Intelligence System is a parallel and interdependent process to the system's definition. However, a conceptual framework should be in place.

Key benefits:

- The concept of the Technology Intelligence system meets the company's **requirements** and can be adapted by necessity.
- The solution is **practical**.
- If existing Technology Intelligence concepts are adaptable, resources for design can be **economized**.

Implementation is an important topic in this book, and some insight could be gained during action research and the validation cases. Certainly it is generally true that implementation of a TI system depends on system design/definition. However, also the contrary is true, which let us say that there is a strong interdependency between the design/definition process and the implementation process. However, as described in Principle 3 there is a fundamental difference between system design and definition. While "design" refers to conceptual ideas, this has to happen before implementation. In turn, "defining" the details of the system is a parallel process to its realization, which is implementation in a narrow view. At the same time, this is a participative learning and adapting (=interdependent) process. Considering this parallel to the definition process, the implementation process can also be seen as answering the questions that emerge from detailing the elements of the holistic TI system.

Principle 7: Roles during Definition and Implementation

The definition and implementation of a Technology Intelligence System is a participative process which requires expert and political "champions."

Key benefits:

- **Expert** champions assure a sound system design/definition.
- **Political** champions assure implementation of the system.

- The **combination** of expert and political champions assure design/definition and implementation of a high quality Technology Intelligence system.

Basic conditions for a successful definition and implementation of a Technology Intelligence system are TI management competencies and "political power" within the organization. The parallel to innovation champions is manifest: there is a need for an "**expert**" in the field of Technology Intelligence and a "**sponsor**" of the definition and implementation project. In smaller companies these two characteristics are probably unified in one single person. Thus, responsible people who may be responsible for definition and implementation of a TI system, and who present either one or both "champion" characteristics are:

- The ***CTO***: Recently an increasing number of companies – large and small – have begun to define and assign the Chief Technology Officer (CTO) function. Tschirky (1998b: 372) gives an overview of the main tasks that a CTO should fulfill: Technology Intelligence is a major topic.
- The ***Head of R&D***: In companies with no CTO, the Head of R&D often completes tasks that are similar to CTO tasks.
- The ***CEO***: The smaller the company is, the more the CEO is actively involved in organizational questions. Since in some companies the CEO is a "client" and "worker" of Technology Intelligence at the same time, it may be reasonable that she or he lead this project.
- A ***member of a planning staff***: Responsibility for definition and implementation of new management systems is often delegated to members of a (strategic) planning staff which is attached to the CTO, CEO or Head of R&D. In this case, the staff member is typically the expert, the CTO, CEO or Head of R&D the sponsor.
- An ***external consultant***: If competencies for such a project are not available internally, an external consultant with specific knowledge and experience in the field of Technology Intelligence may be considered. The champion roles are similar as mentioned above. It must be said that it is highly recommended to adopt a participative approach when building a TI system. Therefore, an external consultant is to be considered as a coach, not a project leader for the definition and implementation process.

Definition and implementation of a Technology Intelligence system should be a **participative** process. Therefore, in addition to the TI project champion(s), other people are involved already before the system is run definitively. Above all, it is highly recommended to involve all top management during definition of the TI mission and goals. Then, potential TI workers should also be involved as soon as possible for two reasons: firstly, acceptance is higher if concerned people are involved during definition, and therefore are also responsible for the system, secondly, since these people are experts, they are most competent to define information sources and to assign roles. However, this involvement should not drift to a request program.

Principle 8: Gatekeeper Approach

Technology Intelligence activities should be participative, and activities should be based on the most competent people. These people should build a formal and coordinated network.

Key benefits:

- Reverting to the most competent people guarantees **optimal quality** of the Technology Intelligence results.
- Sharing activities among existing employees **optimizes resource** expenditures for Technology Intelligence.
- The participative approach makes the TI process (i.e. decision-making) **fast**.
- The participative approach is positive for **organizational learning**.
- In addition, participation basically increases **acceptance**.
- Formalizing and coordinating the gatekeeper network makes activities **transparent**, and therefore improves the system's **quality**. Coordination implies management of TI.

This principle gives the primary suggestion of how to design the Technology Intelligence system in a technology-based SME. This principle evokes four main questions: Who are the most competent people and what are their tasks? What does participative approach mean? What is a formal network? What does coordination mean?

A precondition for assignment of responsibilities for action in the TI system is to set TI mission and goals and if possible, set concrete observation fields (issues). Based on these issues, the **most competent people** should be assigned to them. "Competent" in this context means having well-recognized background and experience in the technological field or related areas, including knowledge of how technologies are currently or potentially applicable to products and processes within the company or by other organizations, and how these technologies may be substituted. Ideally, such an expert presents gatekeeper skills. Some problems will probably emerge from assigning such experts:

- There is an expert for an issue, but she/he does not present gatekeeper skills. In this case, trade-offs have to be made. On the one hand, "imperfect" gatekeepers can be trained (for example communication training). On the other hand, the second-best expert may be a better gatekeeper, and therefore more valuable for the Technology Intelligence system.
- There is an expert for an issue with gatekeeper skills, but she/he does not have enough time for TI activities. Since most competent people normally operate at full capacity, this is a common problem. Observation in this study shows that after an initial (and natural) resistance of the assigned gatekeepers, they can be convinced that "being a gatekeeper in the TI system" does not necessarily differ from their daily work – and therefore does not use a lot of additional time – and that this is a possibility to give their competencies and merits a face, which may be favorable for their career. However, attention should be paid that work quality, neither for TI activities nor for daily work, does not decrease because of time demands. Again, if the expert really is unable to work as a gatekeeper, the second-best expert may be more valuable as a gatekeeper.
- There is no internal competence for a defined issue. Two options are suggested in this case. The preferred option is to build this competence internally. This means that an employee who works in a neighboring field or who is a generalist, for example a young scientist, needs to become an expert. The second option is to revert to external experts. However, this solution is sub-optimal because despite the external nature of Technology Intelligence, the value-creating process is an internal affair. Therefore an internal person should assure that the quality of the external's "intelligence" is not less than the "real" gatekeeper's.

The **tasks** of a gatekeeper depend on the TI mission and goals. In general, she/he is responsible for observation/analysis of her/his issue and communication of facts and trends within this issue. It is recommended to define the specifications. Some examples are: What are the information sources, what external networks should be cultivated?

How should information be analyzed, what are analysis criteria? How should intelligence be disseminated, how does the generated intelligence "reach" decisions, respectively how is intelligence accessible? At least by defining these specifications, the gatekeeper will notice that her/his TI activities match well with her/his daily activities.

A participative approach means that Technology Intelligence is not a secret or private affair, but that TI clients and TI workers are actively involved in the TI process. Thus, "participative" stands for a common company effort. The most valuable benefits of a participative approach are speed, quality and organizational learning. Decisions can be made very quickly if the distance between information collection and decision-making is short. This is the case with a participative approach. For example, if a multidisciplinary and multi-hierarchical panel meets regularly, for example monthly, previously gathered information can be analyzed, communicated and applied at the same time. Because participants have various backgrounds, for example technical, financial etc., the quality of analysis should be high. Such participation leads to organizational learning and organizational intelligence at the same time. It is important for the gatekeepers' motivation that the system, i.e. the gatekeeper network, is transparent and that results, i.e. decisions based on the generated intelligence, are visible.

Transparency has already been mentioned several times as an important characteristic of a Technology Intelligence system. In order to maximize this transparency it is recommended to **formalize** the gatekeeper network. In addition to transparency, which increases organizational learning, formalization makes direct and fast communication possible and avoids redundancies. Formalization of the network presents three major interests: Firstly, a gatekeeper who is aware of being part of a network can anticipate the system, which means that she/he can consciously push information and can propose improvements. Secondly, she/he is more motivated because her/his work is at the forefront. Thirdly, since the gatekeeper's work is at the forefront, time spent is justified. Nevertheless, as mentioned before, the network should not be over-formalized.

Coordination is important for the Technology Intelligence system's management. The importance during design/definition and implementation has been discussed before. During "system operations" this coordination should be responsible for appropriate direction and development of the system. This coordinating task can be assigned directly to the CTO (or equivalent) or to an attached staff member. In order to assure TI management, the principal tasks are: Having the overview of people involved and of activities, motivating and supervising gatekeepers, organizing and leading meetings, coordinating the budget, questioning the system permanently and adapting/improving elements of the system.

Principle 9: Tool Deployment

> For information collection, analysis and communication purposes, an SME should revert to Technology Intelligence tools (methods and infrastructure). However, deployment of "technical" tools should be minimized in order to respond to the SME's resource restriction.

Key benefit:

- Appropriate tools support the TI process and therefore make the process more **effective**.

Deployment of TI tools (methods and infrastructure) is probably one of the first questions which emerge when an SME decides to implement a TI system. Therefore and in order to be thorough, an answer to this question is given by this ninth principle. However, experience during this underlying research has shown that **TI tool deployment is not a major concern** in reality. This strengthens the point of view that TI methods and infrastructure are not solutions but support Technology Intelligence.

Considering **TI methods**, it is highly recommended to avoid or, at least, simplify **quantitative "technical"** TI methods, such as S-curve analysis, frequency analysis, bibliometrics etc. Accurate application of such methods is knowledge and time-intensive, and therefore is unaffordable for SMEs. Simplifying such methods could be a solution if the result is still significant. Accordingly, these results should be interpreted having in mind this simplification. For example, if an S-curve is based on assumptions rather than on facts, the result is also an assumption and does not precisely reflect reality. In turn, application of **qualitative "soft"** methods, such as multidisciplinary and inter-subjective opinion forming, is highly recommended. Such methods are applicable during all stages in the TI process. Mostly, emphasis is on communication (also in stages other than dissemination of intelligence). Reverting to qualitative curves, portfolios etc. assures a common understanding and therefore improves discussion and analysis quality. In a conclusion, TI methods should be used in SMEs with a focus on qualitative rather than quantitative methods.

Until now **TI infrastructure** has been considered as Information Technology infrastructure. Independent of size, the company should consider IT use for database purposes. This does not cause a particular problem to most SMEs. For communication purposes, IT should be handled with care. Communication by means of IT makes sense whenever this is the only option or it is more efficient and effective than direct communication. Otherwise, the latter should be preferred. Thus, TI infrastructure is also something other than Information Technology. Whatever is favorable for communication, for example short walking-distances, informal meeting points etc., can be considered as TI infrastructure, and should therefore be managed.

Principle 10: Performance Measurement

> The performance of the Technology Intelligence System should not be measured with quantitative indicators but on a qualitative basis. However, a TI system should be budgeted and controlled.

Key benefits:

- Budgeting and performance measurement enables **justification** of the Technology Intelligence system.
- Measuring performance also indicates **effectivity** and **efficiency** of the TI system.

The basic question behind "performance measurement" is if output (decisions that contribute to company success) justifies input (resources). Consequently this question makes sense. However, studied SMEs do not allow insight into institutionalized performance measurement because this was never a real topic. Nevertheless some insight about input and output could be gained.

A Technology Intelligence system needs resources (see Principle 2). Money and time spent can be measured or, at least estimated. Experience at Straumann showed that investment and running costs listed in Table 3.4 are quite realistic. Investment costs are negligible. For running costs, in turn, a 500 employee company has to count on a **budget** of US$ 70,000 to US$ 200,000. It is recommended to account and control this budget under "Technology Intelligence" in order gain insight into real money spent for TI activities. To dilute information about money spent for Technology Intelligence would hinder the TI system from having an official status, and therefore would decrease acceptance. **Human resource** spending has already been discussed as being difficult. In particular adopting the gatekeeper approach, estimating additional time spent for Technology Intelligence in relation to time spent for daily business is delicate and cannot be more than indicative. Even more difficult is estimating the TI system's **output**: How much money could be saved with a decision that was based on gatekeeper information? How much money could be earned from a business that emerged from TI insight?, etc. Balancing budgeted and effective costs and time spent on an inter-subjective basis with contributions to overall company success is suggested. If top management feels that output justifies input, the TI system can basically continue. If not, an analysis of why top management thinks that there is a mismatch between costs and benefit has to be done. Consequences could be termination or result in the improvement of the TI system.

Measuring the **effectivity** and **efficiency** of TI activities is recommended to be on a qualitative basis. For example benchmarking with comparable companies (for example with companies observed in this book) gives insight into their TI activities, input and output etc. This provides an idea about whether or not the company is doing the right thing (effectivity). Details cannot be compared by benchmarking because each TI system will differ from one to the other company. Within a company, effectivity and efficiency of different gatekeepers can be measured with a subjective estimation (for example by the network coordinator) or by some indicators. However, energy for exploring different issues may be comparable, and simply counting contacted people does not give any indication about information quality.

In summary, such as for the whole TI system, adapting performance measurement to the organizational characteristics and company culture is suggested. Again, since SMEs tend to be culture-oriented, performance measurement on a qualitative basis is recommended.

9 Summary and Outlook

This book has shown how technology-based SMEs could be organized to take advantage of relevant information about technological facts and trends in the company's environment. The pertinent question was: "How could a technology-based SME design and implement a Technology Intelligence system?" The practitioner's voice and the analysis of the state-of-the-art in Technology Intelligence literature revealed the absence of an answer to this questions and therefore showed gaps which this book aims to fill.

Action research in one technology-based SME allowed the study of and generated new elements for a solution for a Technology Intelligence system. In a second stage, these elements were validated horizontally in comparable technology-based SMEs, and the overall solution was validated vertically through a comparison of TI practice in start-ups and large, multinational companies. Based on the generated and validated solution and elements for a Technology Intelligence system in technology-based SMEs, a new set of management principles could be presented in order to achieve a contribution towards closing the gaps in theory and practice.

This solution for Technology Intelligence system is surely no guarantee for sustainable success in technology-based SMEs; This simply is not possible because of the uncertainty of the future, the diversity of companies and the multitude of general success or failure factors. However, by consideration of the presented management principles the probability of being aware of the future is certainly improved, and therefore is promising for general company success.

If this book stimulates some technology-based SMEs to design and implement a Technology Intelligence system, and if this book is a valuable contribution to them, the author has achieved the main goal.

However, some new challenges and issues for further examination in the field of Technology Intelligence have emerged during the writing of this book:

- **Broader validation of the generated elements**: The generated elements, in particular the Opportunity Landscape and the Screening Process, are based on action research in one single technology-based SME, and each of them is validated in another technology-based SME. Therefore, the empirical basis is, by the nature of action research, narrow. Implementing these elements on a broader basis would give deeper insight into strengths and weaknesses, and would allow variations to be tested. In addition to this broader horizontal validation, validating the generated elements vertically in large organizations, i.e. to scale up elements from a "small" solution to large companies, would be beneficial for both science and management.
- **Alternatives to the generated elements**: For the same reason as described above, the generated and discussed elements present *one* well considered and validated solution for Technology Intelligence in technology-based SMEs. Gaining insight into alternatively generated elements would strengthen and complete the management principles.
- **Broader validation of management principles:** In addition to the generated elements, which represent a possible solution for a TI system in technology-based SMEs, further validation of the management principles would be of scientific and practical interest. Hence, each management principle could be a hypothesis for empirical research.
- **Broader insight into interaction of Technology Intelligence with other management processes**: It is practically impossible to set clear limits between different management issues (for example Technology Intelligence and Knowledge Management) and different management processes (for example intelligence process and organizational learning); they are nested and interrelated. However, for some processes such as Technology Roadmapping, resource allocation etc., Technology Intelligence can be considered a supporting and mostly parallel activity. A detailed examination of the interaction with and the impact on most current management processes of Technology Intelligence would upgrade the holistic understanding of an Integrated Technology Management approach.

- **Performance measurement of Technology Intelligence**: Despite some insight into performance measurement in Management Principle 10, a sound and generally applicable performance measurement system for Technology Intelligence could not be defined, nor did former studies challenge this "hot" topic consistently. However, since this is probably the first question top management will ask, a deeper understanding of various measurement options should be examined. In addition to a stronger acceptance, measurement factors allow the examination and definition of success factors of Technology Intelligence.
- **Interaction between industrial and governmental Technology Intelligence activities**: The popular vision of an inter-company or an industry-wide organization of Technology Intelligence for SMEs did not find approval in industry. Nevertheless, learning from governmental intelligence programs and considering them more than a simple information source would be an asset for SMEs, and vice versa. Examining needs for intelligence and potential contributions from both government and SMEs, and how these needs and contributions could be coordinated would be of economic interest.

Even though the topic of this book is Technology Intelligence and the research objects are technology-based SMEs, insight into challenges and issues in management disciplines that are related to Technology Intelligence could be gained. This revealed other new challenges and issues for further examination. Some of them are:

- **Adaptability of management principles for other management disciplines**: The presented set of new management principles emerged primarily from practical experience gained during this research study. However, it seems that almost all principles would be of interest for design and implementation of other management issues, for example Technology Roadmapping. Adopting these principles in management practice, for example by consultants, and/or management research, and reporting on success and failures of such an adoption would be of interest to technology-based SMEs.
- **Knowledge Management in technology-based SMEs**: This study revealed that communication and knowledge sharing is crucial for the success of Technology Intelligence. Since there is still little insight into Knowledge Management in technology-based SMEs, more research needs to be done. In particular the extent to which

insight from studies in large companies could be valid for small companies (downsizing/selecting) should be examined.

- **Resource management in technology-based SMEs**: In general, resources are supposed to be restricted in SMEs. One solution in this book was, for example, to access lacking knowledge resources by networking with external experts. More SME specific concepts for innovative access to resources, in particular knowledge and finances, and concepts for appropriate allocation of resources, for example a dynamic model, should be explored.
- **More implementation stories**: Implementation of management concepts is a moderately explored field. The major challenge is to compare different implementation strategies. It seems to be difficult to make a direct comparison between different implementation strategies, for example by implementing the same concept in a different manner in different companies, because implementation ideally runs in parallel to concept definition. Therefore, we need more individual implementation stories in order to learn from successes and failures.

These issues are challenges to both management practitioners and management scientists. If next to the inspiration for Technology Intelligence in technology-based SMEs this book motivates practitioners and scientists to challenge these new issues, another main goal of the author is achieved.

Bibliography

Abell, D.F. (1999). Competing Today While Preparing for Tomorrow, in: *Sloan Management Review*, vol. 40 (3), pp. 73–81.

Abernathy, W. & Utterback, J.M. (1978). Pattern of Industrial Innovation, in: *Technology Review*, vol. 80 (7), pp. 40–7.

Aguilar, F.J. (1967). *Scanning the Business Environment*. New York, Macmillan.

Allen, T.J. (1986). Managing the Flow of Technology: *Technology Transfer and Dissemination of Technology Information within R&D Organization*, 3rd ed., Cambridge, MIT Press.

Ansoff, H. I. (1980). Strategic Issue Management, in: *Strategic Management Journal*, vol. 1, 131–48.

Argyris, C., Putman, R. & McLain-Smith, D. (1987). *Action Science*. San Francisco, Jossey-Bass Publishers.

Arvanitis, S. (1997). *Innovation und Unternehmenserfolg in der Schweizer Industrie – Eine empirische Untersuchung anhand von Firmendaten*. Final report, National fund project 12-32443.91. Zürich, Swiss Institute for Business Cycle Research.

Ashton, W.B., Johnson, A.K. & Stacey, G.S. (1994). Monitoring Science and Technology for Competitive Advantage, in: *Competitive Intelligence Review*, vol. 5 (1), pp. 5–16.

Ashton, W.B., Kinzey, B.R. & Gunn, M.E. (1991). A Structured Approach for Monitoring Science and Technology Developments, in: *International Journal of Technology Management*, vol. 6 (1/2), pp. 91–111.

Ashton, W.B. & Klavans, R.A. (1997). An Introduction to Technical Intelligence in Business, in: Ashton, W.B. & Klavans, R.A. (eds.), *Keeping Abreast of Science and Technology: Technical Intelligence in Business*. Columbus, OH, Batelle Press, pp. 5–22.

Ashton, W.B. & Stacey, G. S. (1995). Technical Intelligence in Business: Understanding Technology Threats and Opportunities, in: *International Journal of Technology Management*, vol. 10 (1), pp. 79–104.

Auster, E. & Choo, C.W. (1994). How Senior Managers Acquire and Use Information in Environmental Scanning, in: *Information Processing & Management*, vol. 30 (5), pp. 607–18.

Baisch, F. (2000). *Implementierung von Früherkennungssystemen in Unternehmen*. Köln, Josef Eul.

Barnett, H.G. (1953). *Innovation: The Basis of Cultural Change*. New York, McGraw-Hill.

Beal, R.M. (2000). Competing Effectively: Environmental Scanning, Competitive Strategy, and Organizational Performance in Small Manufacturing Firms, in: *Journal of Small Business Management*. January 2000, pp. 27–47.

Becker, J. (1995). Den Austausch von Wissen fördern, in: *Gablers Magazin*, vol. 3, pp. 16–19.

Berry, M.M.J. (1996). Technical Entrepreneurship, Strategic Awareness and Corporate Transformation in Small High-Tech Firms, in: *Technovation*, vol. 16 (9), pp. 487–98.

Berry, M.M.J. & Taggart, J.H. (1998). Combining Technology and Corporate Strategy in Small High Tech Firms, in: *Research Policy*, vol. 26 (7/8), pp. 883–95.

Bethke, H. & Lang, H.-C. (1998). Elemente des Technologie-Managements bei Novartis Pharma, in: Tschirky, H. & Koruna, S. (eds.), *Technologie-Management: Idee und Praxis*. Zürich, Industrielle Organisation, pp. 698–719.

Beyers, W.B. & Nelson, pp.B. (1998). *The Economic Impact of Technology-based Industries in Washington State in 1997*. A report prepared for the Technology Alliance, Seattle, WA.

Bleicher, K. (1992). *Konzept Integriertes Management*. Frankfurt, Campus.

Boisot, M.H. (1983). Convergence Revisited: The Codification and Diffusion of Knowledge in a British and a Japanese Firm, in: *Journal of Management Studies*, vol. 20 (1), pp. 159–90.

Boisot, M.H. (1998). *Knowledge Assets: Securing Competitive Advantage in the Information Economy*. Oxford, Oxford University Press.

Brenner, M.S. (1996). Technology Scouting and Technology Intelligence, in: *Competitive Intelligence Review*, vol. 7 (3), pp. 20–7.

Bridge, J. & Peel, M.J. (1999). Research Note: A Study of Computer Usage and Strategic Planning in the SME Sector, in: *International Small Business Journal*, vol. 17 (4), pp. 82–7.

Bryant, pp.J., Coleman, J.C. & Krol, T.F. (1997). Organizing a Competitive Technical Intelligence Group, in: Ashton, W.B. & Klavans, R.A. (eds.), *Keeping Abreast of Science and Technology: Technical Intelligence in Business*. Columbus, OH, Batelle Press.

Chabot, C. (1995). *Defining High Technology*. STS Publication of Stanford University.

Chen, J.-C. & Williams, B.C. (1993). The Impact of Electronic Data Interchange (EDI) on SMEs: Summary of Eight British Case Studies, in: *Journal of Small Business Management*, vol. 36 (4), pp.68–73.

Chetty, S. (1996). The Case Study Method for Research in Small- and Medium-sized Firms, in: *International Small Business Journal*, vol. 15 (1), pp. 73–85.

Choo, C.W. (1998). *Information Management for the Intelligent Organization: The Art of Scanning the Environment*. Medford, New Jersey, Information Today.

Chrobok, R. (1998). Wissensmanagment, in: *Zeitschrift für Organisation*, (3), pp. 184–5.

Clemens, R., Günterberg, B., Hauser, H.-E. & Kayser, G. (1997). *Unternehmensgrössenstatistik 1997/98: Daten und Fakten*. Study of the Institut für Mittelstandsforschung Bonn.

Cobbenhagen, J. (2000). *Successful Innovation: Towards a New Theory for the Management of Small and Medium-sized Enterprises*. Cheltenham, Edward Elgar Publishing.

Cohen, W.M. & Levinthal, D.A. (1990). Absorptive Capacity: A New Perspective on Learning and Innovation, in: *Administrative Science Quarterly*, vol. 35 (1), pp. 128–52.

Cook, S.D.N. & Yanow, D. (1996). Culture and Organizational Learning, in: Cohen, M. D. & Sproull, L. S. (eds.), *Organizational Learning*. Thousand Oaks, Sage Publications.

Cunningham, J. B. (1993). *Action Research and Organizational Development*. London, Praeger Publishers.

Curran, J., Jarvis, R., Blackburn, R.A. & Black, S. (1993). Networks and Small Firms: Constructs, Methodological Strategies and Some Findings, in: *International Small Business Journal*, vol. 11 (2), pp. 13–25.

Daenzer, W.F. (1976). *Systems Engineering: Leitfaden zur methodischen Durchführung umfangreicher Planungsvorhaben*. Köln, Hahnstein.

Daft, R.L. & Lengel, R.H. (1986). Organizational Information Requirements, Media Richness and Structural Design, in: *Management Science*, vol. 32 (5), pp. 554–70.

Dankbaar, B. (1996). The Management of Technology in Technology-Contingent SMEs, in: Cannell, W. & Dankbaar, B. (eds.), Technology Management and Public Policy in the European Union, pp. 103–26.

Davis, S. & Botkin, J. (1994). The Coming of Knowledge-Based Business, in: *Harvard Business Review*, vol. 72 (5), pp. 165–70.

Davison, L. (2001). Measuring Competitive Intelligence Effectiveness: Insights from the Advertising Industry, in: *Competitive Intelligence Review*, vol. 12 (4), pp. 25–38.

Dedijer, S. & Jéquier, N. (1987). Intelligence for Economic Development: An Inquiry Into the Role of the Knowledge Society. Oxford, Berg.

Dodgson, M. & Rothwell, R. (1991). Technology Strategies in Small Firms, in: *Journal of General Management*, vol. 17 (1), pp. 45–55.

Dou, H. (1995). *Veille Technologique et Compétitivité*. Paris, Dunod.

Drucker, pp. (1985). The Discipline of Innovation, in: *Harvard Business Review*, vol. 64 (3), pp. 67–72.

Ebadi, Y.M. & Utterback, J.M. (1984). The Effects of Communication on Technological Innovation, in: *Management Science*, vol. 30 (5), pp. 572–85.

Eger, M.C. (1995). *You Can't Measure TI Success? Think Again!*, SCIP Annual Conference Presentation, Phoenix, AZ.

Ergas, H. (1987). Does Technology Policy Matter?, in: Guile, B.R. & Brooks, H. (eds.), *Technology and Global Industry*. Washington, National Academy Press, pp. 191–245.

Escher, J.-Ph. (2001). The Process of External Technology Deployment: Part of the Technology Marketing Process, in: *Proceedings of the 4th PICMET Conference*, Portland, Oregon.

European Commission (1999). Achtung! Die Definition der KMU ändert sich nicht!, in: *Euro-Info – Das Bulletin der Unternehmenspolitik*, no. 1.

European Commission (2000). *The European Observatory for SMEs: Sixth Report, Executive Summary*. Luxembourg.

Ewald, A. (1989). *Organisation des Strategischen Technologie-Managements*. Berlin, Erich Schmidt.

Fann, G.L. & Smeltzer, L.R. (1989). The Use of Information from and about Competitors in Small Business Management, in: *Entrepreneurship Theory and Practice*, Summer 1989, pp. 35–46.

Foster, F. (1999). Justifying Knowledge Management Investments, in: *Knowledge and Process Management*, vol. 6 (3), pp. 154–7.

Foster, R. (1986). *Innovation – The Attacker's Advantage*. London, Macmillan Press.

Fuld, L.M. (1995). *The New Competitor Intelligence: The Complete Resource for Finding, Analyzing and Using Information About Your Competitors*. New York, John Wiley & Sons.

Gälweiler, A. (1987). *Strategische Unternehmensführung*. Frankfurt, Campus.
Gerybadze, A. (1994). Technology Forecasting as a Process of Organisational Intelligence, in: *R&D Management*, vol. 24 (2), pp. 131–40.
Gibb, A. & Davies, L. (1990). In Pursuit of Frameworks for the Development of Growth Models of the Small Business, in: *International Small Business Journal*, vol. 9 (1), pp. 15–31.
Gilad, B. & Gilad, T. (1988). *The Business Intelligence System: A New Tool for Competitive Advantage*. New York, NY, Amacom.
Gomez, pp. (1996). Von der ganzeitlichen Unternehmensentwicklung zum Wissensmanagement, in: *Wissensmanagement: Wettbewerbsvorteile durch den gezielten Einsatz des Wissenskapitals*, seminar at the St. Gallen University.
Greenwood, D. J. & Levin, M. (1998). *Introduction to Action Research. Social Research for Social Change*. Thousand Oaks, CA, Sage Publications.
Guimaraes, T. & Armstrong, C. (1998). Exploring the Relations Between Competitive Intelligence, IS Support and Business Change, in: *Competitive Intelligence Review*, vol. 9 (3), pp. 45–54.
Gupta, A.K. & Wilemon, D. (1996). Changing Patterns in Industrial R&D Management, in: *Journal of Product Innovation Management*, vol. 13 (6), pp. 497–511.
Hadjimanolis, A. (2000). A Resource-Based View of Innovativeness in Small Firms, in: *Technology Analysis & Strategic Management*, vol. 12 (2), pp. 263–81.
Hall, C. (2001). The Intelligence Puzzle, in: *Competitive Intelligence Review*, vol. 12 (4), pp. 3–14.
Hambrick, D.C. (1982). Environmental Scanning and Organizational Strategy, in: *Strategic Management Journal*, vol. 3 (2), pp. 159–74.
Hauschildt, J. (1993). *Innovationsmanagement*. München, Franz Vahlen.
Hauser, H.-E. (2000). *SMEs in Germany: Facts and Figures 2000*. Study of the Institut für Mittelstandsforschung Bonn.
Hassid, L., Jacques-Gustave, pp. & Moinet, N. (1997). *Les PME face au défi de l'intelligence économique*. Paris, Dunod.
Hedin, H. (1993). Business Intelligence: Systematized Intelligence Activities in Ten Multinational Companies, in: *Journal of AGSI*, vol. 2 (4), pp. 126–36.
Hoffman, K., Parejo, M., Bessant, J. & Perren, L. (1998). Small Firms, R&D, Technology and Innovation in the UK: A Literature Review, in: *Technovation*, vol. 18 (1), pp. 39–55.
Hohhof, B. (1997). Computer Support Systems for Scientific and Technical Intelligence, in: Ashton, W.B., Klavans, R.A. (eds.), *Keeping Abreast of Science and Technology. Technical Intelligence in Business*. Columbus, OH, Batelle Press, pp. 259–79.
Japan Small and Medium Enterprise Agency (2001). *White Paper on Small and Medium Enterprises in Japan: A Wake-Up Call to Small Business, Building a Self-Sustaining Enterprise*. Tokyo.
Jakobiak, F. & Dou, H. (1992). De l'information documentaire à la veille technologique pour l'entreprise, in: Desvals, H. & Dou, H. (eds.), *La veille technologique: l'information scientifique, technique et industrielle*. Paris, Dunod.
Jennings, D.F. & Lumpkin, J.R. (1992). Insights Between Environmental Scanning and Porter's Generic Strategies: An Empirical Analysis, in: *Journal of Management*, vol. 18 (4), pp. 791–803.

Jung, H.H. & Tschirky, H. (2002). *Technology Control in Management Control Systems*, submitted for: the Academy of Management Conference 2002. Denver, Submission no. 10242.

Junnarkar, B. (1997). Leveraging Collective Intellect by Building Organizational Capabilities, in: *Expert Systems With Applications*, vol. 13 (1), pp. 29–40.

Kahaner, L. (1997). *Competitive Intelligence: How to Gather, Analyze, and Use Information to Move Your Business to the Top*. New York, Touchstone.

Keller, G. (1997). *Erstellung eines Informationsquellenmix zur Beschaffung von strategischen Informationen für die Technologiefrühaufklärung*. Project work at the ETH Zürich. Zürich, Center for Enterprise Science.

Kobe, C. (2001). *Integration der Technologiebeobachtung in die Frühphase von Innovationsprojekten*. St. Gallen, Doctoral dissertation at the University of St. Gallen.

Kodama, F. (1991). *Analyzing Japanese High Technologies: The Techno-paradigm Shift*. London, Pinter Publishers.

Kohler, O. (1994). *Technologie-Management in schweizerischen kleinen und mittelgrossen Unternehmen*. Zürich, Dissertation ETH no. 10477.

Koruna, S. (2001). *Wissensmanagement*. Internal Report, ETH Zürich.

Krystek, U. & Müller-Stewens, G. (1993). *Frühaufklärung für Unternehmen: Identifikation und Handhabung zukünftiger Chancen und Bedrohungen*. Stuttgart, Schäffer-Poeschel.

Kubicek, H. (1975). *Empirische Organisationsforschung: Konzeption und Methodik*. Stuttgart, Poeschel.

Lang, H.-C. (1998). *Technology Intelligence: Ihre Gestaltung in Abhängigkeit der Wettbewerbssituation*. Zürich, Industrielle Organisation.

Lewin, K. (1946). Action Research and Minority Problems, in: *Journal of Social Issues*, vol. 2 (4), pp. 34–46.

Lichtenthaler, E. (2000). *Organisation der Technology Intelligence: eine empirische Untersuchung in technologieintensiven, international tätigen Grossunternehmen*. Zürich, Dissertation ETH no. 13787.

Lichtenthaler, E. & Tschirky, H. (2001). *Structures of Technology Intellligence and Innovation Systems*. Working-Paper ETH Zürich.

Lybaert, S. (1998). The Information Use in SME: Its Importance and Some Elements of Influence, in: *Small Business Economics*, vol. 10 (2), pp. 171–91.

Lyles, M.A. & Baird, I.S. & Orris, J.B. & Kuratko, D.F. (1993). Formalized Planning in Small Business: Increasing Strategic Choices, in: *Journal of Small Business Management*, vol. 31 (2), pp. 38–50.

Major, E.J. & Cordey-Hayes, M. (2000). Engaging the Business Support Network to Give SMEs the Benefit of Foresight, in: *Technovation*, vol. 20 (11), pp. 589–602.

Marquis, D.G. (1969). *Successful Industrial Innovations – A Study of Factors Underlying Innovation in Selected Firms*. Washington, National Science Foundation (NSF 69–17).

Matthews, C.H. & Scott, S.G. (1995). Uncertainty and Planning in Small and Entrepreneurial Firms: An Empirical Assessment, in: *Journal of Small Business Management*, vol. 33 (4), pp. 34–52.

McDonald, D.W. & Richardson, J.L. (1997). Designing and Implementing Technology Intelligence Systems, in: Ashton, W.B. & Klavans, R.A. (eds.), *Keeping Abreast of Science and Technology: Technical Intelligence in Business*. Columbus, OH, Batelle Press, pp. 123–55.

Minder, S. (2001). *Wissensmanagement in KMU: Beitrag zur Ideengenerierung im Innovationsprozess*. St. Gallen, Verlag KMU HSG.

Mintzberg, H. & Lampel, J. (1999). Reflecting on the Strategy Process, in: *Sloan Management Review*, vol. 40 (3), pp. 21–30.

Mitterdorfer, D. (2001). *Modellierung unternehmensspezifischer Innovations-Prozessmodelle. Zürich*, Dissertation ETH no. 14212.

Mohan-Neill, S.I. (1995). The Influence of a Firm's Age and Size on its Environmental Scanning Activities, in: *Journal of Small Business Management*, vol. 33 (4), pp. 10–21.

Myers, L.A. (1983). Information Systems in Research and Development: The Technological Gatekeeper Reconsidered, in: *R&D Managemen*t, vol. 13 (4), pp. 199–206.

Nefiodow, L.A. (1990). *Der fünfte Kondratieff: Strategien zum Strukturwandel in Wirtschaft und Gesellschaft*. Frankfurt, Frankfurter Allgemeine.

Nonaka, I. & Takeuchi, H. (1995): *The Knowledge-Creating Company: How Japanese Companies Create the Dynamics of Innovation*. New York, Oxford University Press.

OECD (1997). *Science, Technology and Industry: Scoreboard of Indicators*. Paris, OECD.

Pavitt, K. (1980). *Technical Innovation and British Economic Performance*. London, MacMillan Press.

Peiffer, S. (1992). *Technologie-Frühaufklärung: Identifikation und Bewertung zukünftiger Technologien in der strategischen Unternehmensplanung*. Hamburg, S&W Steuer- und Wirtschaftsverlag.

Polanyi, M. (1966). *The Tacit Dimension*. London, Routledge and Kegan Paul.

Porter, A.L., Roper, A.T., Mason, T.W., Rossini, F.A., Banks, J. & Wiederholt, B.J. (1991). *Forecasting and Management of Technology*. New York, John Wiley & Sons.

Porter, M.E. (1980). *Competitive Strategy: Techniques for Analyzing Industries and Competitors*. New York, Free Press.

Porter, M.E. (1985). *Competitive Advantage: Creating and Sustaining Superior Performance*. New York, Free Press.

Prescott, J.E. & Smith, D.C. (1989). A Framework for the Design and Implementation of Competitive Intelligence Systems, in: Snow, C.C. (ed.), *Strategy, Organization Design, and Human Resource Management*. Greenwich, CT, Jai Press.

Probst, G.J.B., Raub, S. & Romhardt, K. (1999). *Wissen managen*. Frankfurt/Wiesbaden, Frankfurter Allgemeine, Gabler.

Raisinghani, M.S. (2000). Knowledge Management: A Cognitive Perspective on Business and Education, in: *American Business Review*, vol. 18 (2), pp. 105–12.

Rapoport, A. (1988). *Allgemeine Systemtheorie, wesentliche Begriffe und Anwendungen*. Darmstadt, Darmstadt Blätter.

Reger, G. & Blind, K. & Cuhls K. & Kolo, C. (1998). *Technology Foresight in Enterprises. Main Results of an International Study by the Fraunhofer Institute for Systems and Innovation Research (ISI) and the Department of R&D Management*. University of Stuttgart.

Ribault, J.M., Martinet, B. & Lebidois, D. (1991). *Le management des technologies*. Paris, Les éditions d'organisation.

Sanchez, R., Heene, A. & Thomas, H. (1996). *Dynamics of Competence-Based Competition: Theory and Practice in the New Strategic Management*. Oxford, New York: Elsevier.

Savioz, P. (2001). Intelligence zur Entscheidungsunterstützung in High-Risk-Projekten, in: Gassman, O., Kobe, C. & Voit E. (eds.), *High-Risk-Projekte: Quantensprünge in der Entwicklung erfolgreich managen*. Berlin, Springer, pp. 279–300.

Savioz, P., Lichtenthaler, E., Birkenmeier, B., Brodbeck, H. (2002). Organisation der frühen Phasen des radikalen Innovationsprozesses, in: *Die Unternehmung*, vol. 6.

Schmookler, J. (1966). *Invention and Economic Growth*. Cambridge, Harvard University Press.

Schneider, U. (1996). *Wissensmanagement: die Aktivierung des intellektuellen Kapitals*. Frankfurt, Frankfurter Allgemeine.

Schumpeter, J.A. (1939). *Business Cycle – A Theoretical, Historical and Statistical Analysis of Capitalist Process*, vol. 1. New York, McGraw-Hill.

Scott, G.M. (2000). Critical Technology Management Issues of New Product Development in High-Tech Companies, in: *Journal of Product Innovation Management*, vol. 17 (1), pp. 57–77.

Scott, G.M. (2001). Strategic Planning for Technology Products, in: *R&D Management*, vol. 31 (1), pp. 15–26.

Senge, pp. (1990). *The Fifth Discipline: The Art and Practice of the Learning Organization*. New York, Currency/Doubleday.

Simons, R. (1995). *Levers of Control; How Managers Use Innovative Control Systems to Derive Strategic Renewal*. Boston, Harvard Business School Press.

Smeltzer, L.R., Fann, G.L. & Nikolaisen, V.N. (1988). Environmental Scanning Practices in Small Business, in: *Journal of Small Business Management*, vol. 26 (3), pp. 55–62.

Sparrow, J. (2001). Knowledge Management in Small Firms, in: *Knowledge and Process Management*, vol. 8 (1), pp. 3–16.

Sveiby, K.E. (1997). *The New Organizational Wealth: Managing and Measuring Knowledge-based Assets*. San Francisco, Berrett-Koehler Publishers.

TopNano21 (2001). *Innovationscheck Nanotechnology: A Tool to Identify and Evaluate Enterprise Specific Application Fields within Nanotechnology*. First interim report, KTI-Project 5049.1. Zürich, EHT-Center for Enterprise Science.

Tschirky, H. (1998a). Technologie-Management: Schliessung der Lücke zwischen Management-Theorie und Technologie-Realität, in: Tschirky, H. & Koruna, S. (eds.), *Technologie-Management: Idee und Praxis*. Zürich, Industrielle Organisation, pp. 1–32.

Tschirky, H. (1998b). Konzept und Aufgaben des Integrierten Technologie-Managements, in: Tschirky, H. & Koruna, S. (eds.), *Technologie-Management: Idee und Praxis*. Zürich, Industrielle Organisation, pp. 193–394.

Tschirky, H. (2000). On the Path of Enterprise Science?: An Approach to Establishing the Correspondence of Theory and Reality in Technology-intensive Companies, in: *International Journal of Technology Management*, vol. 20 (3/4), pp. 405–28.

Tschirky, H. (2002). Wake-up Call for General Management: It's Technology Time, in: EITM (eds.), *Bringing Technology into the Boardroom*. Hampshire, Palgrave.

Twiss, B.C. (1992). *Forecasting for Technologists and Engineers: A Practical Guide for Better Decisions*. London, Peter Peregrinus.

Udwadia, F.E. (1990). Creativity and Innovation in Organizations – Two Models and Managerial Implications, in: *Technological Forecasting and Social Change*, vol. 38 (1), pp. 65–80.

Ulrich, H. & Probst, G. (1988). *Anleitung zum ganzheitlichen Denken und Handeln – ein Brevier für Führungskräfte*. Bern, Haupt.

Utterback, J. M. & Brown, J. W. (1972). Monitoring for Technological Opportunities, in: *Business Horizons*, vol. 15 (10), pp. 5–15.

Vahs, D. & Burmester, R. (1999). *Innovationsmanagement, von der Produktidee zur erfolgreichen Vermarktung*. Stuttgart, Schäffer-Poeschel.

Von Zedtwitz, M. (1999). *Managing Interfaces in International R&D*. St. Gallen, Dissertation no. 2315.

Welsh, J.A. & White, J.F. (1980). A Small Business is not a Little Big Business, in: *Harvard Business Review*, vol. 59 (4), pp. 18–32.

Wright, pp., Pringle, C. & Kroll, M. (1992). *Strategic Management Text and Cases*. Needham Heights, Allyn and Bacon.

Yasai-Ardekani, M. & Nystrom, pp. (1996). Designs for Environmental Scanning Systems: Tests of a Contingency Theory, in: *Management Science*, vol. 42 (2), pp. 187–204.

Yin, R. K. (1988). *Case Study Research: Design and Methods*. Newbury Park, CA, Sage Publications.

Index

BOWLING GREEN STATE UNIVERSITY
DISCARDED
LIBRARY